Proceedings of the 6th Annual Workshop

POLARIZED
POSITRON
2011

Proceedings of the 6th Annual Workshop

POLARIZED POSITRON 2011

Beijing, China   28 – 30 August 2011

Editor

Wei Gai

Argonne National Lab, USA

World Scientific

NEW JERSEY • LONDON • SINGAPORE • BEIJING • SHANGHAI • HONG KONG • TAIPEI • CHENNAI

Published by

World Scientific Publishing Co. Pte. Ltd.
5 Toh Tuck Link, Singapore 596224
USA office: 27 Warren Street, Suite 401-402, Hackensack, NJ 07601
UK office: 57 Shelton Street, Covent Garden, London WC2H 9HE

British Library Cataloguing-in-Publication Data
A catalogue record for this book is available from the British Library.

POLARIZED POSITRON 2011
Proceedings of the 6th Annual Workshop

ISBN 978-981-4401-03-6

Printed in Singapore by World Scientific Printers.

PREFACE

Positron, the antiparticle of electron which is more connected to our everyday life, is much less familiar even to scientists, except those who work on high energy physics, since positron reveals itself only through either weak interaction or pair production from a sufficiently energetic photon.

Positrons, as source particles for high energy physics experiments, in electron positron colliders, for example, should be produced with sufficient quantity and required quality from a well designed and constructed positron source. How to design and how to construct a positron source in a high energy accelerator is a big issue to positron source community, although this community is relatively quite small.

As for future electron positron colliders, such ILC and CLIC, the effort towards positron sources demonstrates itself through a workshop series, so-called "POSIPOL", a workshop started from 2006 at CERN. From August 28-30, 2011, POSIPOL 2011 was held in IHEP, Beijing. As the workshop chairman, again, I was impressed by the charm, the technical difficulty, and the expertise of the positron source, and again, I was aware the great value of this small positron source community. During the workshop, Dr. Wei Gai, one of the leading scientists for ILC positron source working group from Argonne National Laboratory and the editor of this book, proposed to have a book published on the subject of positron sources for linear colliders with the aim of providing an ensemble of related scientific knowledge. This proposal was strongly echoed by the participants, and many of them served as authors for the articles in this book.

Asked by the editor, I write these words with honor and pleasure, but most importantly, with the great respect for those who work tirelessly on the subject year by year, decade by decade, towards a common goal of realizing future linear colliders in the world.

Prof. Jie Gao
Institute of High Energy Physics
Beijing, China.
Asia Linear Collider Steering Committee Chairman
2012-05-07

CONTENTS

PHYSICS APPLICATIONS OF POLARIZED POSITRONS

S. Riemann

DESY Zeuthen, Platanenallee 6,
D-15738 Zeuthen, Germany
E-mail: sabine.riemann@desy.de

With the LHC a new era of measurements at the energy frontier has started, and exciting new discoveries are expected. However, also measurements at the precision frontier will be necessary to fully understand the underlying physics model. The programme for the e^+e^- collider projects ILC and CLIC is focused on precision tests of the Standard Model and new physics beyond it at the TeV scale. Polarized positron beams play a crucial role in these analyses. Here, the advantages as well as the requirements using also polarized positron beams for measurements at e^+e^- colliders are discussed.

Keywords: Linear collider; positron source; polarized positrons; physics with polarized beams.

1. Introduction

So far, the particle physics experiments have confirmed the Standard Model (SM) with excellent precision. Neither significant deviations from the SM predictions nor new physics phenomena have been obtained. Based on the global analysis of the measurements it is expected that the SM Higgs boson has a mass of $\mathcal{O}(100)$ GeV. The fundamental question whether the Higgs boson exists will be answered soon by the measurements at the LHC, and the experiments are well prepared to discover and probe new physics beyond the SM. But the full understanding of phenomena obtained at the LHC is only possible if complementary measurements from lepton colliders are available. The precise knowledge of type, energy and helicity of the interacting particles allows to test theoretical models at the level of quantum corrections up to higher orders. The microscopic world of electroweak interactions is not left-right symmetric and so are new phenomena suggested by various extensions of the SM. The chiral structure of interactions can be analyzed best using high-energy lepton colliders with polarized beams. However, the production of an intense, highly polarized electron beam with high energy is simple in comparison to the generation of the corresponding polarized positron beam. But the flexibility and the substantial advantages justify the effort necessary to create the polarized positron beam.

In this paper important features of measurements at e^+e^- colliders with polarized beams are discussed. Section 2 presents few selected examples for precision physics with polarized beams. In subsections 2.1–2.3 the basics of measurements with polarized beams are introduced. The experimental requirements to utilize polarized positron beams are described in section 3. Section 4 summarizes.

2. Physics with Polarized Positrons

The era of precision electroweak measurements[2] at high energies was based on experiments at the Large Electron Positron Collider (LEP) at CERN and at the SLAC Linear Collider (SLC). The Standard Model has been confirmed with extremely high precision, up to loop corrections. Its parameters have been determined and the mass of the SM Higgs boson has been predicted. One of the important SM parameters that describe the electroweak symmetry breaking is the weak mixing angle, $\sin^2 \theta_W$. The measurement of this observable was performed by the four LEP collaborations, ALEPH, DELPHI, L3 and OPAL, and by the SLD collaboration; the results and details can be found in reference[2]. It was impressive to see that the SLD collaboration achieved a slightly more precise measurement of this parameter than the four LEP collaborations combined although the latter obtained a more than 30 times higher number of Z bosons created in e^+e^- collisions. The crucial point was the polarized electron beam which increased the sensitivity to the left-right asymmetry of the Z boson coupling to fermions. If SLD would have used also polarized positrons a further reduction of the uncertainty by a factor of about two would have been possible.

This simple Gedankenexperiment demonstrates the potential of polarized beams in high energy particle physics experiments. The precise test of the SM at high energies as well as the understanding of new phenomena benefit substantially if electron and positron beams are polarized. A comprehensive overview of physics with both beams – electrons and positrons – polarized is given in reference[1]. Here, some of the basics are emphasized.

First, few remarks about the requirements for measurements at the precision frontier. Future lepton colliders have to complement and to attend the physics goals achieved with the LHC. This implies physics at center-of-mass energies between 200 GeV and 1 (3) TeV. Two projects are under development: the International Linear Collider (ILC)[3] with energies between 200 GeV and 1 TeV and the possibility to run also at the Z boson resonance, $\sqrt{s} = 91.2\,\text{GeV}$, and the Compact Linear Collider (CLIC)[4] foreseen for energies up to 3 TeV. To interpret the results and to examine the SM and possible extensions, the precision of measurements must be better than the size of higher order corrections to the observables. With other words: Only high intensities (combined with a highly sophisticated detector) allow to detect the huge number of events for all interesting processes which is necessary to perform measurements with uncertainties at and below the percent-level. However, it is not at all easy to produce a beam with the required high luminosity. Since the cross sections in lepton colliders fall as $\sigma \sim 1/E^2$, the increase of energy by a factor f has to be compensated by a factor f^2 for the luminosity to keep the number of events almost constant. Further, the stability of energy and luminosity must be very high – below 0.1% for the ILC – and the precise measurements of energy and luminosity must be possible. Similar requirements exist for the beam polarization. As shown in the SLD experiment at SLC, electron beam polarization of 80% is possible and measurable with a precision of 0.5%.[2] Further improvements are possible at the ILC.[5] In the following features of precision measurements using polarized beams are discussed.

2.1. Fermion-Pair Production in the s-Channel

Some important advantages of physics with colliding polarized beams can be explained best for the fermion-pair production process. Photon and Z boson are spin-1 particles, and in the SM they are exchanged in this process, $e^+e^- \to Z, \gamma \to f\bar{f}$. The Feynman diagram in lowest order is shown in Figure 1. For longitudinally polarized beams, the cross section can be

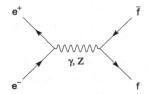

Fig. 1. The Feynman diagram in lowest order for the fermion-pair production; in the SM, photon and Z boson are exchanged (J=1).

written as

$$\sigma_{\mathcal{P}_{e^-}\mathcal{P}_{e^+}} = \frac{1}{4}\left[(1-\mathcal{P}_{e^-})(1+\mathcal{P}_{e^+})\sigma_{\text{LR}} + (1+\mathcal{P}_{e^-})(1-\mathcal{P}_{e^+})\sigma_{\text{RL}} \right.$$
$$\left. + (1-\mathcal{P}_{e^-})(1-\mathcal{P}_{e^+})\sigma_{\text{LL}} + (1+\mathcal{P}_{e^-})(1+\mathcal{P}_{e^+})\sigma_{\text{RR}}\right], \quad (1)$$

with the electron beam polarization $\mathcal{P}_{e^-}$ and the positron beam polarization $\mathcal{P}_{e^+}$. σ_{LR} denotes the cross section if the electron beam is 100% left-handed polarized ($\mathcal{P}_{e^-} = -1$), and the positron beam 100% right-handed polarized ($\mathcal{P}_{e^+} = +1$). The other cross sections, σ_{RL}, σ_{LL} and σ_{RR}, are defined correspondingly. Since the exchange of the spin-1 particles, photon and Z boson, in the fermion-pair production is only possible for $J = 1$, the cross sections σ_{RR} and σ_{LL} are zero in the SM[a]. Figure 2 shows the possible combinations of electrons and positrons with helicities ± 1. It is not excluded that further – yet unknown – particles contribute either to processes with $J = 1$ or $J = 0$. If these particles are not too heavy they can be studied by precise measurements of the process $\text{e}^+\text{e}^- \to \text{f}\bar{\text{f}}$.

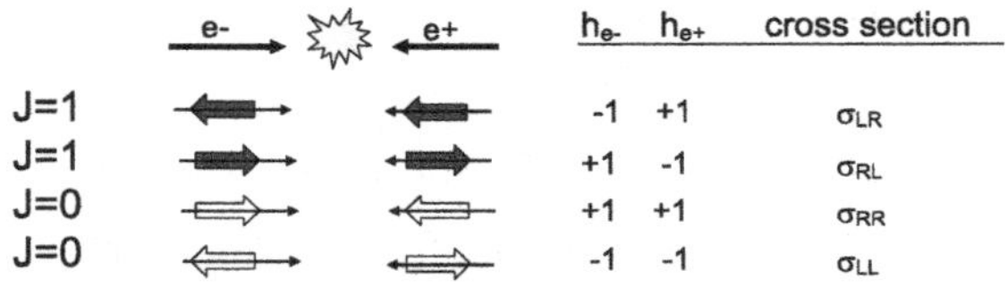

	e-	e+	h_{e^-}	h_{e^+}	cross section
J=1	←	←	-1	+1	σ_{LR}
J=1	→	→	+1	-1	σ_{RL}
J=0	→	←	+1	+1	σ_{RR}
J=0	←	→	-1	-1	σ_{LL}

Fig. 2. Helicity combinations in collisions of electrons and positrons and the corresponding contributions to the cross section.

The cross section (1) can be expressed with

$$\sigma_{i,j} = \frac{1}{4}\sigma_u\left[1 - \mathcal{P}_{e^+}\mathcal{P}_{e^-} + A_{\text{LR}}\left(-\mathcal{P}_{e^+} + \mathcal{P}_{e^-}\right)\right], \quad (2)$$

where A_{LR} is the left-right asymmetry caused by the different coupling strength of Z bosons to left- an right-handed fermions, and the indices i, j describe the sign of the polarization: σ_{-+}, σ_{+-}, σ_{++}, σ_{--}. Taking into account beams with realistic polarization, $|\mathcal{P}| < 1$, the measured cross

---

[a]The cross section for the exchange of Higgs bosons ($J = 0$) yields only tiny contributions and is neglected.

sections for the different helicity combinations are

$$\sigma_{-+} = \frac{1}{4}\sigma_u \left[1 + |P_{e+}P_{e-}| + A_{LR}\left(+|P_{e+}| + |P_{e-}|\right)\right] \tag{3}$$

$$\sigma_{+-} = \frac{1}{4}\sigma_u \left[1 + |P_{e+}P_{e-}| + A_{LR}\left(-|P_{e+}| - |P_{e-}|\right)\right]$$

$$\sigma_{++} = \frac{1}{4}\sigma_u \left[1 - |P_{e+}P_{e-}| + A_{LR}\left(-|P_{e+}| + |P_{e-}|\right)\right] \tag{4}$$

$$\sigma_{--} = \frac{1}{4}\sigma_u \left[1 - |P_{e+}P_{e-}| + A_{LR}\left(+|P_{e+}| - |P_{e-}|\right)\right]$$

where σ_u denotes the cross section with unpolarized beams. The cross sections σ_{++} and σ_{--} $(J = 0)$ are zero for $\mathcal{P}_{e-} = \mathcal{P}_{e+} = \pm 1$, in contrast to $\sigma_{+-} \neq 0$ and $\sigma_{-+} \neq 0$ for $\mathcal{P}_{e-} = -\mathcal{P}_{e+} = \pm 1$ $(J = 1)$. It is easy to see that in case of unpolarized electron and positron beams half of the collisions is spent for helicity combinations that yield $\sigma = 0$. Figure 2 illustrates the combinations and resulting cross section contributions.

If the electron beam is 100% longitudinally polarized, but the positron beam is unpolarized, one half of the measurements is performed with the orientation $\mathcal{P}_{e-} = +1$, the other half with $\mathcal{P}_{e-} = -1$. Also in this case initial state helicity combinations occur that do not contribute to the cross section. Hence, only half of the possible helicity combinations yields $\sigma \neq 0$.

However, if both beams are 100% polarized and $\mathcal{P}_{e-} = -\mathcal{P}_{e+}$, all possible combinations of initial state helicity amplitudes contribute to the cross section measurement and the luminosity is enhanced compared to the case of unpolarized beams. Figure 3 demonstrates these options.

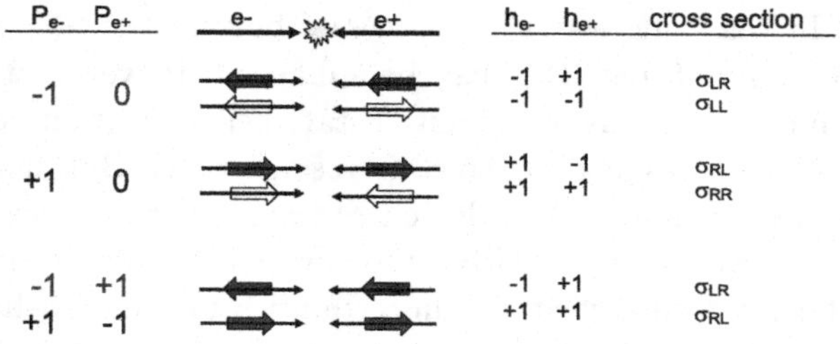

Fig. 3. Helicity combinations in collisions of a longitudinally polarized electron and unpolarized positron beam (upper part) and in collisions with both beams polarized. The corresponding helicities of the initial state particles as well as the contributions to the cross section are shown.

2.2. Cross Sections

It is an important result, that the effective luminosity can be substantially enhanced if both beams are polarized. The unpolarized cross section, σ_u, is given by the sum

$$\sigma_u = \frac{1}{4} \left(\sigma_{+-} + \sigma_{-+} + \sigma_{--} + \sigma_{++} \right) . \tag{5}$$

Using unpolarized beams, σ_u is measured obtaining the number N_u of events for the integrated luminosity $\mathcal{L}$,

$$\sigma_u = \frac{N_u}{\mathcal{L}} . \tag{6}$$

If the electron beam is 100% polarized but the positron beam unpolarized, and the luminosity is equally distributed to collisions with $\mathcal{P}_{e^-} = -1$ and $\mathcal{P}_{e^-} = +1$, one finds

$$\sigma_u = \frac{\sigma_{+0} + \sigma_{-0}}{2} = \frac{N_{+0} + N_{-0}}{\mathcal{L}/2} = \frac{N_u}{\mathcal{L}} . \tag{7}$$

If also the positron beam is polarized, the unpolarized cross section is

$$\sigma_u = \frac{\sigma_{-+} + \sigma_{+-}}{2} = \frac{N_{+-} + N_{-+}}{\mathcal{L}/2} = \frac{N_u}{\left(1 - \mathcal{P}_{e^+}\mathcal{P}_{e^-}\right)\mathcal{L}} . \tag{8}$$

The luminosity is effectively enhanced,

$$\mathcal{L}_{\text{eff}} = \left(1 - \mathcal{P}_{e^+}\mathcal{P}_{e^-}\right)\mathcal{L} , \tag{9}$$

resulting in a smaller statistical error of the measurement. With positron beam polarization of $|\mathcal{P}_{e^+}| = 0.4$ (0.6), the effective luminosity can be increased by about 30% (50%) having an electron beam polarization of $|\mathcal{P}_{e^-}| = 0.8$. In the same way, also processes beyond the SM, e.g. due to the exchange of spin-0 particles, can be enhanced. However, in that case also runs with combinations of the initial state helicities are necessary that are 'inefficient' with respect to the SM cross sections. But the flexibility to chose the desired initial state helicities improves the precision of SM measurements as well as the sensitivity to new phenomena beyond the SM.

It must be mentioned that the uncertainties for the left-handed and right-handed cross section measurements, $\delta\sigma_{+-}$, $\delta\sigma_{-+}$, include also the error of the polarization measurement. For $\delta\mathcal{P}_{e^+}/\mathcal{P}_{e^+} = \delta\mathcal{P}_{e^-}/\mathcal{P}_{e^-} = \delta\mathcal{P}/\mathcal{P}$ the additional error contribution due to the beam polarization measurement is

$$\frac{\delta\sigma_{ij}}{\sigma_{ij}} = \frac{\delta\mathcal{P}}{\mathcal{P}} \sqrt{2\mathcal{P}_{e^+}^2 \mathcal{P}_{e^-}^2 + A_{\text{LR}}^2 \left(\mathcal{P}_{e^+}^2 + \mathcal{P}_{e^-}^2\right)} , \tag{10}$$

which is unimportant for small relative polarization errors and small A_{LR}. However, for high luminosities larger $1\,\mathrm{ab}^{-1}$ and $\delta\mathcal{P}/\mathcal{P} = 0.25\%$, this contribution can approach the magnitude of the statistical error of the cross section measurement. The corresponding contribution to the uncertainty of the unpolarized cross section is

$$\frac{\delta\sigma_u}{\sigma_u} = \frac{\mathcal{P}_{e+}\mathcal{P}_{e-}}{1 - \mathcal{P}_{e+}\mathcal{P}_{e-}} \sqrt{\left(\frac{\delta\mathcal{P}_{e+}}{\mathcal{P}_{e+}}\right)^2 + \left(\frac{\delta\mathcal{P}_{e-}}{\mathcal{P}_{e-}}\right)^2}. \tag{11}$$

and increases slightly the uncertainty of the measurement. The knowledge of the contributions (10) and (11) is important for precision measurements with high luminosities and high beam polarizations. Large errors on the polarization measurement could limit the precision to measure unpolarized quantities, or the right-handed and left-handed cross-sections, σ_{LR} and σ_{RL}, correspondingly.

2.3. *Left-Right Asymmetry*

The left-right asymmetry A_{LR} is an important observable to measure the left- and right-handed coupling of bosons to fermions. It is defined as

$$A_{LR} = \frac{\sigma_{LR} - \sigma_{LR}}{\sigma_{LR} + \sigma_{LR}}. \tag{12}$$

Since in realistic beams $|\mathcal{P}| < 1$, A_{LR} is derived from measurements by

$$A_{LR} = \frac{\sigma_{-+} - \sigma_{+-}}{\sigma_{-+} + \sigma_{+-}} = \frac{A_{LR}^{\mathrm{meas}}}{\langle\mathcal{P}_{\mathrm{eff}}\rangle} \tag{13}$$

with the effective polarization, $\mathcal{P}_{\mathrm{eff}}$:

$$\mathcal{P}_{\mathrm{eff}} = \frac{-\mathcal{P}_{e-} + \mathcal{P}_{e-}}{1 - \mathcal{P}_{e-}\mathcal{P}_{e+}} \tag{14}$$

The effective polarization is larger than the individual $e^{\pm}$ beam polarizations; 80% polarization of the electron beam are increased to an effective polarization of almost 95% using a 60% polarized positron beam. Because of error propagation the uncertainty of the effective polarization is substantially decreased. Assuming that the relative error for polarization measurement of the electron and positron beam is $\delta\mathcal{P}_{e+}/\mathcal{P}_{e+} = \delta\mathcal{P}_{e-}/\mathcal{P}_{e-} = \delta\mathcal{P}/\mathcal{P}$, the uncertainty of the effective polarization yields

$$\frac{\delta\mathcal{P}_{\mathrm{eff}}}{\mathcal{P}_{\mathrm{eff}}} = \frac{\delta\mathcal{P}}{\mathcal{P}} \frac{\sqrt{\left(1 - \mathcal{P}_{e+}^2\right)^2 \mathcal{P}_{e-}^2 + \left(1 - \mathcal{P}_{e-}^2\right)^2 \mathcal{P}_{e+}^2}}{\left(\mathcal{P}_{e+} + \mathcal{P}_{e-}\right)\left(1 + \mathcal{P}_{e+}\mathcal{P}_{e-}\right)}. \tag{15}$$

Assuming 80% (90%) electron polarization, and an uncertainty of the polarization measurement of $\delta\mathcal{P}/\mathcal{P} = 0.25\%$,[5] the error of the effective polarization is reduced by a factor 3.7 if the positron beam is 60% polarized. This fact is important for precise A_{LR} measurements with large integrated luminosity. In this case the error contribution due to the polarization uncertainty could dominate the total error, δA_{LR}, given by

$$\delta A_{\mathrm{LR}} = \sqrt{\frac{1 - \mathcal{P}_{\mathrm{eff}}^2 A_{\mathrm{LR}}}{\mathcal{P}_{\mathrm{eff}} N} + A_{\mathrm{LR}}^2 \left(\frac{\delta\mathcal{P}_{\mathrm{eff}}}{\mathcal{P}_{\mathrm{eff}}}\right)^2}. \tag{16}$$

2.4. *u,t-Channel Processes*

In sections 2.1–2.3 some basic advantages are discussed for s-channel processes. Without going into detail it should be mentioned that the search for new phenomena benefits from polarized positrons also if u- and t-channel processes are considered. In u- and t-channel processes the helicity of the particle's final state is directly coupled to the helicity of the initial state fermion, it does not depend on the helicity of the second incoming beam particle. This gives a direct access to the helicity of the exchanged particle and allows an enhancement or suppression of specific processes. An example is the production of single W bosons, $e^+e^- \to W\,e\,\nu$, which is one basic process to study CP violation. For more details and examples, in particular the sensitivity to supersymmetric phenomena, the interested reader is strongly encouraged to consult reference[1].

2.5. *W⁺W⁻ Pair Processes*

The precise measurement of the Three-Gauge-Boson-Coupling (TGC) in the process $e^+e^- \to Z, \gamma \to W^+W^-$ allows a test of the weak gauge structure as described by the SM, and it is very sensitive to new physics scenarios. Since the SM defines the TGC, deviations of precision measurements from the SM prediction are hints to new phenomena. To select this process with high efficiency, the contribution of the neutrino exchange in the t-channel, $e^+e^- \to \nu \to W^+W^-$ (see Figure 4), is suppressed using a polarized electron beam. A further improvement is possible with polarized positrons in addition to polarized electrons.

2.6. *Higgs Factory*

The Higgs boson is a scalar particle which can be produced in e^+e^- annihilation by the Higgsstrahlung or boson fusion (see Figure 5). The dominating process is determined by the Higgs mass which is not yet known.

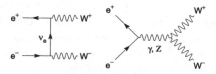

Fig. 4. Feynman diagrams for the process $^+e^- \to W^+W^-$. Only the right diagram is important to measure TGC.

In case of a light Higgs boson as suggested by the electroweak precision measurements at LEP and SLD,[2] the Higgsstrahlung is dominating. With polarized positrons the Higgs production can be enhanced by a factor $(1 - \mathcal{P}_{e+}\mathcal{P}_{e-})$. If the Higgs boson is heavy, it is produced via WW fusion, $e^+e^- \to \nu_e\bar{\nu}_e H$. This process can be enhanced (or suppressed) by the factor $(1 + \mathcal{P}_{e+})(1 - \mathcal{P}_{e-})$ choosing the proper sign of the $e^\pm$ polarizations. For $(\mathcal{P}_{e-}, \mathcal{P}_{e+}) = (-80\%, +60\%)$, the WW fusion process is enhanced by a factor of 2.88 in comparison to unpolarized beams.

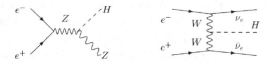

Fig. 5. Feynman diagrams for Higgs production processes: Higgsstrahlung process (left) and WW boson fusion (right).

2.7. GigaZ Option

Electroweak precision measurements at the Z resonance were performed by the experiments ALEPH, DELPHI, L3 and OPAL at LEP and SLD at SLC. Taking into account the results for the top quark mass and the W boson mass, the SM has been confirmed at the one-loop level of quantum corrections. The results of LEP and SLD are in good agreement, however, the A_{LR} measurement at SLD results in values for the weak mixing angle or correspondingly for the Z boson couplings to fermions that are slightly different from that determined by the LEP experiments. Running at the Z resonance again by utilizing polarized $e^\pm$ beams and a much higher luminosity would substantially improve the accuracy of electroweak measurements. This option is called GigaZ since the luminosity available at the ILC allows to produce and record about 10^9 Z bosons within few months of running. High-luminosity measurements at the Z resonance combined with updated precise results for the W boson mass, the top-quark mass, and hopefully the Higgs boson mass allow excellent consistency tests of the SM and provide a high sensitivity to models beyond the SM. This would also to test whether the slightly differing values for observables measured at LEP and SLD is a fluctuation as assumed so far, or whether it is due to a certain yet unknown phenomenon. At GigaZ a relative precision of less than 5×10^{-5} can be achieved for the effective weak mixing angle – more than 10 times better than the values achieved with LEP/SLD measurements. This allows precise conclusions on new physics models, e.g. supersymmetry. The GigaZ option requires very precise measurements of energy, luminosity and beam polarization. If both beams are polarized, the Blondel scheme[6] can be applied to determine the beam polarization and A_{LR} simultaneously with highest precision.[7]

2.8. *Transversely Polarized Beams*

Finally it must be mentioned that also collision of transversely polarized beams are interesting. They allow access to helicity correlations, CP violating effects and new phenomena like extra dimension.[8] The contribution to the differential cross section due to transverse polarization is

$$\frac{\mathrm{d}\sigma}{\mathrm{d}\Omega} \sim \mathcal{P}_\perp^{\mathrm{e}^+} \mathcal{P}_\perp^{\mathrm{e}^-} \sin\theta \cos 2\phi \,, \tag{17}$$

which is zero if one of the colliding beams is unpolarized. New physics phenomena yield additional terms resulting in substantially modified differential cross sections. For example, extra dimensions differential cross sections measured for transversely as well as longitudinally polarized beams with angular distributions typical for special classes of models. Hence, physics runs with transversely polarized beams will help to distinguish between models and to resolve ambiguities.

3. Requirements for Physics

Polarized positron and polarized electron beams offer the best flexibility and an improved sensitivity to fulfill the physics programme for future high energy lepton colliders. Unfortunately, it is quite difficult to produce an intense polarized positron beam for a high energy linear collider. The ILC baseline design proposes to generate the positrons using photons created in an undulator passed by a high energy electron beam.[9] Since the photon yield in a helical undulator is higher up to a factor 2 than in a planar undulator, the ILC positron source design is based on a helical undulator. The photons generated in a helical undulator are circularly polarized. If they hit the positron production target, longitudinally polarized electron-positron pairs are created. The scheme has been tested successfully in the proof-of-principle experiment E-166 at SLAC.[10]

Using a helical undulator, the ILC will provide a polarized positron beam. The degree of polarization is determined by the parameters of the undulator and the source design. The opening angle of the photon beam decreases with the electron energy, $\propto 1/\gamma$. The polarization of the photons is distributed such that a collimation of the photon beam increases its average polarization. The loss of intensity has to be compensated using a longer undulator. For more details see the references[11,12].

In order to exploit the positron and electron beam polarization for physics measurements, the degree of polarization must be kept up to the interaction point. Therefore spin rotation systems upstream the damping ring rotate the particle spins from the longitudinal to the vertical direction, parallel (or anti-parallel) to the magnetic field in the damping ring. Downstream the turnaround the spins are rotated back to the longitudinal direction so that the beams have the desired polarization at the IP.

The electron and positron beam polarization is measured at the IP using Compton polarimeters. To meet the high precision requirements, the relative uncertainty of the polarization measurement must be at the level of (few) per-mille.

One important issue must be mentioned: The direction of the helical undulator winding determines the orientation of the photon polarization and therefore the sign of the positron polarization. Switching to the opposite orientation of positron beam polarization requires an additional spin-flip equipment. This point is discussed in the subsection 3.3. It should be remarked that in a polarized positron source based on Compton back-scattering the helicity reversal can be easily realized by switching the polarization of the laser light.

3.1. *Polarimetry at the Interaction Point*

The beam polarization at the interaction point is measured using Compton polarimeters. In order to determine the luminosity-weighted longitudinal polarization at the interaction point (IP) at the ILC, one polarimeter is located at the beginning of the Beam Delivery System upstream the IP, the other in the extraction line downstream the IP. The two polarimeters are highly complementary. The upstream polarimeter has a clean environment and a much higher counting rate; the fast polarization measurement is important to detect correlations. The downstream polarimeter measures the polarization of the outgoing beam. Since the background in the downstream polarimeter is high and the beam is disrupting after the IP, the counting rate is substantially smaller than in the upstream polarimeter. But the downstream polarimeter has access to depolarization effects: Due to the small bunch sizes high electromagnetic fields act between the particles in the crossing bunches and induce the radiation of hard photons. The resulting depolarization has to be taken into account to attain the required precision of the polarization measurement. The combination of both polarimeters allows the determination of the luminosity-weighted polarization; cross checks between both polarimeters are possible. Measurements without collisions can be used to control the spin transport through the Beam Delivery System. However, due to the large beam disruption at CLIC the downstream polarimeter will not work with the required precision.

Present studies and test measurements show that at the ILC a precision of $\delta\mathcal{P}/\mathcal{P} \approx 0.25\%$ can be achieved[5] for the longitudinal polarization. For comparison: the precision for the polarization measurement reached with the Compton polarimeter at the SLD experiment was $\delta\mathcal{P}/\mathcal{P} = 0.5\%$. The measurement of the transverse polarization at the IP is under study.

3.2. *Positron Polarimetry at the Source*

Since the production of an intense positron beam needs some effort the degree of polarization should also be measurable at the positron source. At the electron source a Mott polarimeter is used. Due to the design and the parameters of a polarized positron source it is not recommended to apply a Mott polarimeter. Instead, a Bhabha polarimeter located at beam energies of few hundred MeV is a promising option.[13]

3.3. *Frequency of Helicity Reversal*

As discussed in section 2, a substantial enhancement of the effective luminosity is possible with polarized beams. But the increase by the factor $(1 - \mathcal{P}_{e^-}\mathcal{P}_{e^+})$ is only possible in case of an efficient pairing of initial states $(+-)$, $(-+)$. This requires the same helicity reversal frequencies for the electron and the positron beam. The polarization of the electron beam can be flipped easily by reversing the polarity of the laser beam which hits the photocathode. A fast and random flipping between the beam polarization orientations reduces systematic uncertainties substantially (see also reference[2]). The orientation of the positron beam polarization can be reversed using a spin rotator. However, it is impossible to switch the high magnetic field in the spin rotator within very short time, e.g. from train to train as possible for the electron beam. There is no gain for the effective luminosity if the helicity of the positrons is reversed from run to run (or even less often) and the the helicity of the electrons train-by-train. Further, to control systematic effects, a very high long-term stability is necessary. A possible solution of this problem would be to kick the positron beam to parallel spin rotation lines with opposite magnetic fields, similar as suggested in reference[14].

The precision measurements require almost identical intensities and polarizations for the left- and right-handed oriented beams. The measured left-right asymmetry is related to the left-right asymmetry by

$$A_{\mathrm{LR}} = \frac{A_{\mathrm{LR}}^{\mathrm{meas}}}{\langle \mathcal{P}_{\mathrm{eff}} \rangle} . \qquad (18)$$

If the luminosities and degrees of polarization are identical for σ_{-+} and σ_{+-}, one gets

$$A_{\mathrm{LR}} = \frac{N_{-+} - N_{+-}}{N_{-+} - N_{+-}} \frac{1}{\langle \mathcal{P}_{\mathrm{eff}} \rangle} . \qquad (19)$$

Also for fast helicity reversal small differences in luminosity and polarization occur between the running modes $(+-)$ and $(-+)$. They have to be taken into account,

$$A_{\mathrm{LR}} = \frac{A_{\mathrm{LR}}^{\mathrm{meas}}}{\langle \mathcal{P}_{\mathrm{eff}} \rangle} + \frac{1}{\langle \mathcal{P}_{\mathrm{eff}} \rangle} \left[(A_{\mathrm{LR}}^{\mathrm{meas}})^2 A_{\mathcal{P}} + \langle \mathcal{P}_{\mathrm{eff}} \rangle \Delta_{\mathcal{P}} + A_{\mathcal{L}} + \ldots \right] , \qquad (20)$$

where $A_{\mathcal{L}}$ and $A_{\mathcal{P}}$ are the left-right asymmetries of the integrated luminosity and of the beam polarization; the asymmetries of residual background, the center-of-mass energy, detector acceptance and efficiency are not shown in equation (20). The contribution $\Delta_{\mathcal{P}}$ depends on $\Delta \mathcal{P}_{e^+} \mathcal{P}_{e^-} + \Delta \mathcal{P}_{e^-} \mathcal{P}_{e^+}$ with $\Delta \mathcal{P}_e$ as difference between $+$ and $-$ sign of the beam polarization. A slower helicity reversal for the positron than for the electron beam yields different luminosities for the running modes $(+-)$ and $(-+)$, and also the degree of polarization could vary. The resulting corrections in equation (20) could be large. The corrections to A_{LR}, i.e. $A_{\mathcal{L}}$, $A_{\mathcal{P}}$ and $\Delta_{\mathcal{P}}$, must be determined and should be as small as possible. In particular, the uncertainty of $A_{\mathcal{L}}$ and $\Delta_{\mathcal{P}}$ should be at the per-mille level to achieve the desired high precision for A_{LR}.

Detailed studies are ongoing to evaluate the influence of parallel spin rotation lines on the final physics performance with polarized beams, and to find alternative solutions with fast and flexible helicity reversal at the undulator-based positron source.

4. Summary

Precision measurements of SM physics and phenomena beyond the SM can be performed at future linear e^+e^- colliders. They will extend and complement the achievements of the LHC. The best conditions are provided if high luminosity, a wide energy range and polarized beams are available. In particular, the flexible choice of initial state helicities is desired to reveal unexpected phenomena and their nature.

The polarization of both beams, electrons and positrons, affords substantial advantages: The occurrence of desired processes can be enhanced. The effective luminosity for s-channel processes with exchange of spin-1 particles can be increased by the factor $(1 - \mathcal{P}_{e^-}\mathcal{P}_{e^+})$ if the luminosity is equally distributed to running modes with the initial state helicities $(+-)$ and $(-+)$. The uncertainty of the effective polarization is reduced which is important for precision measurements of left-right asymmetries. Among many arguments to have polarized positrons it should be emphasized: If signals from physics beyond the SM are found, a much better distinction between models is possible than with only one polarized beam. For the GigaZ option the electron and the positron beam must be polarized to achieve the required precision for the A_{LR} and polarization measurement. In order to benefit from these advantages, it must be possible to reverse the helicity of positrons as frequent as the helicity of electrons. Hence, for a positron source based on a helical undulator an additional facility is necessary to realize the fast spin flip for the positrons.

Finally, it should be emphasized that a positron source based on a helical undulator will provide a polarized positron beam; the degree of polarization depends strongly on the undulator parameters and the energy of the electrons passing through. One may ask what minimum degree of positron polarization is necessary. Recent ILC studies[1,15] show that for $\mathcal{P}_{e^+} > 30\%$ the physics analyses clearly benefit from polarized electron and positron beams. Of course, a high degree of positron polarization is desired and can be realized by photon beam collimation for the undulator-based source (see also reference[16]). Thus, an excellent feasibility is provided to perform the high energy linear collider physics programme at the precision frontier.

Acknowledgments

I would like to thank my collaborators, in particular Gudrid Moortgat-Pick and Jenny List for discussions about physics with polarized positron beams. I am grateful to Prof. Jie Gao and his local team for organizing this successful POSIPOL 2011 Workshop. I enjoyed the interesting sessions, the fruitful discussions, the social program and the stay in Beijing.

References

1. G. Moortgat-Pick, *et al.*, Phys. Rept. **460**, 131 (2008) [hep-ph/0507011].
2. ALEPH, DELPHI, L3 and OPAL and SLD and LEP Electroweak Working Group and SLD Electroweak Group and SLD Heavy Flavour Group Collaborations, Phys. Rept. **427**, 257 (2006) [hep-ex/0509008].
3. J. Brau, (Ed.) *et al.* [ILC Collaboration], arXiv:0712.1950 [physics.acc-ph];
   G. Aarons *et al.* [ILC Collaboration], arXiv:0709.1893 [hep-ph];
   T. Behnke, (Ed.) *et al.* [ILC Collaboration], arXiv:0712.2356 [physics.ins-det].
4. R.W. Assmann *et al.*, CERN-2000-008; E. Accomando *et al.*, hep-ph/0412251;
   http://project-clic-cdr.web.cern.ch/project-CLIC-CDR/
5. S. Boogert, M. Hildreth, D. Käfer, J. List, K. Mönig, K. C. Moffeit, G. Moortgat-Pick and S. Riemann *et al.*, JINST **4**, P10015 (2009) [arXiv:0904.0122 [physics.ins-det]].
6. A. Blondel, Phys. Lett. B **202**, 145 (1988) [Erratum-ibid. **208**, 531 (1988)].
7. J. Erler J *et al.*, Phys. Lett. B **486**, 125 (2000).
8. T. G. Rizzo, [arXiv:1011.2185 [hep-ph]];
   T. G. Rizzo, JHEP **0308**, 051 (2003) [hep-ph/0306283];
   T. G. Rizzo, JHEP **0302**, 008 (2003) [hep-ph/0211374].
9. V.E. Balakin and A.A. Mikhailichenko, Budker Institute of Nuclear Physics Report No. BINP 79-85, 1979; R.C. Wingerson, Phys. Rev. Lett. **6**, 446 (1961).
10. G. Alexander *et al.*, Nucl. Instrum. Meth. A **610**, 451 (2009) [arXiv:0905.3066 [physics.ins-det]]; G. Alexander *et al.*, Phys. Rev. Lett. **100**, 210801 (2008).
11. A. Ushakov *et al.*, these proceedings, arXiv:1202.0752 [physics.acc-ph].
12. W. Gai, these proceedings.
13. G. Alexander *et al.*, EUROTeV-Report-2008-091.
14. K. Moffeit *et al.*, SLAC-TN-05-045.
15. M. Berggren, arXiv:1007.3019 [hep-ex].
16. F. Staufenbiel *et al.*, these proceedings, arXiv:1202.5987 [physics.acc-ph].

POSITRON SOURCE FOR INTERNATIONAL LINEAR COLLIDER

WEI GAI, WANMING LIU

High Energy Physics Division, Argonne National Lab, USA

In this paper, we presented an overview of the positron source for International Linear Collider (ILC). We started with description of the positron source configuration for ILC RDR baseline and SB2009 baseline followed by the status of critical components of ILC positron source R&D. We also presented some parameters of positron source for both RDR and SB2009 baseline.

1. Introduction

The positron source for International Linear Collider requires a huge number of positrons to be produced and accepted by damping ring [1]. This requirement is far beyond any existing positron source for linear colliders and brings in a lot of challenges to the designing and realization of ILC positron source.

The ILC baseline positron source is an undulator based positron source. The electron main linac beam passes through a long helical undulator to generate a photon beam with energy from few MeV to several hundreds of MeV which then strikes a thin metal target to generate positrons in an electromagnetic shower. The positrons are captured, accelerated, separated from the shower consitituents and unused photon beam and then are transported to the damping ring.

The ILC positron source works in a pulsed mode with a large number of bunches in a pulse at a repetition rate of 5Hz and the repetition rate will be increased to 10Hz for low energy runes as documented recently in SB2009. The 10Hz low energy running scenarios are asserted as a result of insufficient positron yield when the drive beam energy is lower than 150GeV nominal drive beam energy for ILC RDR baseline.

Besides the baseline helical undulator based positron source for ILC, there are other alternative schemes like Laser Compton Scattering based and conventional positron source under R&D in parallel. But here we'll concentrate on the ILC undulator based positron source and discuss about the key elements and parameters.

Lay out as in RDR baseline

In the RDR baseline[1], the positron source subsystem started with a insertion of an over 100m long of undulator at where the main electron beam is of an energy of 150GeV. The 150GeV main electron beam get bended into the undulator where it will lost about 3GeV of energy into photon beam and then bended back into electron main linac to be accelerated or decelerated to its' desired before get into beam delivery system (BDS). The photon beam produced by the 150GeV drive beam in the ~115m long undulator will bombard into a

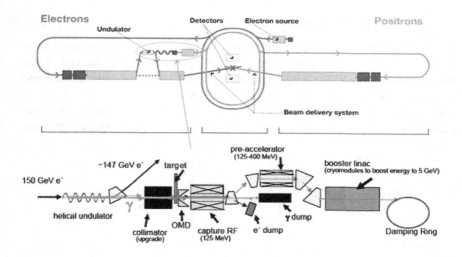

Figure 1. Schematic of ILC RDR and undulator based positron source

thin titanium target (0.4 X0) about 400m downstream from the end of undulator. Following immediately after the conversion target, an optical matching device (OMD) is employed to enhance the capture efficiency. The OMD used in RDR baseline is an AMD with a Bz field of 5T on the surface of target and then adiabatically decreased down 0.5T in 20cm. Passing through the OMD, the positrons will then be accelerated up to 125MeV using normal conducting linacs and then separated from electrons and photons. After separation, the positron beam will then be accelerated up to 400MeV in pre-accelerator and then transported to supper conducting booster linac to boost the energy up to 5GeV followed by a bunch compressing and spin rotation before got injected into damping ring.

The detail of the beam line optics for RDR baseline positron source can be found in [2] and [3]. As described in [2], the ILC RDR positron source beam line is divided into the Positron CAPture (PCAP), the Positron Pre-Accelerator (PPA), the Positron Pre-Accelerator To the Electron main Linac tunnel (PPATEL), the Positron TRANsport (PTRAN), the Positron BooSTeR linac (PBSTR) and the Linac To Ring (LTR). In the PCAP, positron produced in the target will be captured and accelerated up to 125MeV. A systematic simulation of positron source from undulator to 5GeV damping ring entrance can be found in [4].

2. Lay out as in SB2009 proposal

As shown in figure 2, in SB2009 proposal [5], changes related to positron source in SB2009 include: Undulator-based positron source located at the end of the electron Main Linac (250 GeV), in conjunction with a Quarter-wave transformer as capture device; a lower beam-power parameter set with the number of bunches per pulse reduced by a factor of two (nb = 1312), as compared to the nominal RDR parameter set; reduced circumference Damping Rings (~3.2 km)

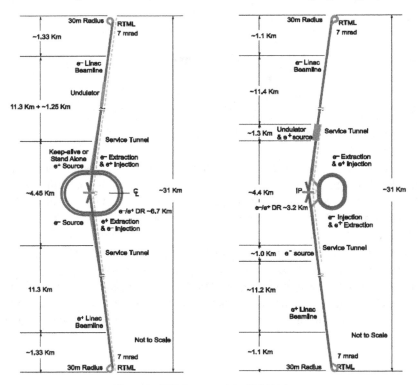

Figure 2. RDR layout and the SB2009 layout

at 5 GeV with a 6 mm bunch length; integration of the positron and electron sources into a common "central region beam tunnel", together with the BDS, resulting in an overall simplification of civil construction in the central region.

The change with most significant impact on positron source is to move the undulator based positron source to the end of electron main linac to lower the cost and simplify the civil construction. As a result of such change, the drive beam energy of the undulator based positron source will not be fixed and a study of the performance under different drive beam energies has to be done to understand the risks and solutions.

3. Critical components of undulator based positron source

As mentioned before, the undulator based positron source is consist of a long undulator, a thin titanium alloy conversion target, OMD, positron capturing RF linacs, transportation beamline optics, booster linacs, spin rotator and bunch compressor. Among those components, undulator, target, OMD and capturing RF linacs are the most unique and critical components for positron source.

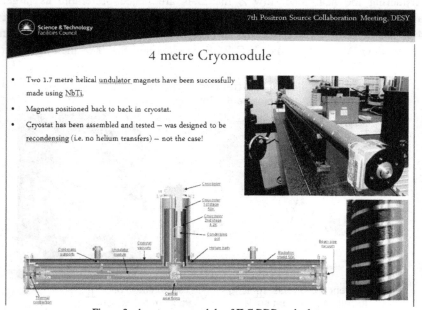

Figure 3. 4 meter cryomodule of ILC RDR undnulator

The undulator determines the spectrum of photon radiated for a given electron drive beam. It has great impact on the positron yield and also the polarization property of resulting positron source. The ILC RDR baseline has chosen a baseline of K=0.92 and λu=1.15cm. As shown in figure 3, a 4 meter long cryo-module with two 1.7m long RDR undulator has been Completed at STFC/RAL/Daresbury[5]. With 137m RDR undulator, when driven with 150GeV drive beam, the RDR can provide a positron yield of 1.5 with a 0.4X0 Titanium target and an AMD with 5T at the surface of target. When a quarter wave transformer is used in place of AMD, 231m long RDR undulator is required in order to achieve the 1.5 positron yield. When the drive beam energy is lower or higher as in SB2009, the positron yield can be higher or lower. The detail will be discussed later.

The conversion target is where the positron produced as a result of pair production in the field of a nucleus in the target. Unalike the conventional positron source where the target needs to be 4-6 radiation length thick, the undulator based positron source requires a target of only 0.4-0.5 radiation length. Since the ILC requires about 5.3×10^{13} positrons per RF pulse, thermal stress in the target has to be kept under control in order to avoid target damage. Low Z materials have in general a higher heat capacity than high Z materials. Considering the heat load, low Z materials are hence preferable as target material. Some studies before shown that at the optimum target thickness of 0.4X0 the yield for a low Z material as Ti is only about 16% below the yield of an equivalent W target[6]. For ILC RDR baseline and SB2009, the target is a

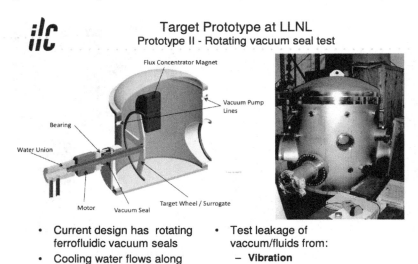

Figure 4. Rotating vacuum seal test at LLNL

titanium wheel with a diameter of 2 meter and a rim of 2cm wide and 1.4cm thick. In order to take away the heat deposition, a cooling system with rotating vacuum seal for the target is been prototyping and testing for leakage of fluid and vacuum from the effect of vibration and magnetic field at LLNL. Figure 4 shows the vacuum seal test setup at LLNL[7].

As shown in the schematic layout, target will be followed by an OMD and will be likely exposed to high magnetic field except for cases using lithium lens. The effect from the eddy currents of moving metal objects in magnetic field has also been studied [8] and experiment has been set up and done data taking (STFC/LLNL)[9]. The experiment setup at STFC is showing in figure 5.

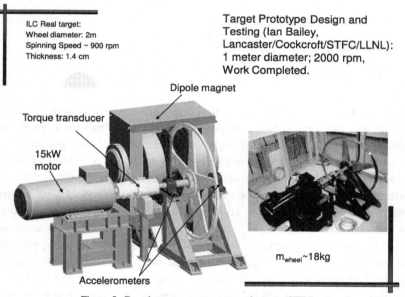

Figure 5. Rotating target protype experiment at STFC

OMD is used to enhance the positron capture efficiency in a positron source and there exist several different type of OMD. Numerical studies have done to compare them under the same conditions, same undulator, target and capturing RF. As shown in table 1, for RDR undulator configuration, both AMD and lithium lens has a capture efficiency of about 30%. But with AMD, the 2 meter diameter target wheel needs to be able rotating at 900RPM in 6T magnetic field which makes it impractical if not impossible. For lithium lens, the performance is about the same of an ideal AMD. But given the beam intensity of ILC positron source, concerns about the survivability of windows need to be answered before it can be seriously considered. To be conservative, the quarter wave transformer is chosen to be used in the RDR baseline and meanwhile, the prototyping of flux concentrator is being carried out at LLNL.

Table 1. Capture efficiency of different OMD

OMD	Capture efficiency
Immersed target, AMD (6T-0.5T in 20 cm)	~30%
Non-immersed target, flux concentrator (0-3.5T in 2cm, 3.5T-0.5T 14cm)	~26%
1/4 wave transformer (1T, 2cm)	~15%
0.5T Back ground solenoid only	~10%
Lithium lens	~29%

Due to the extremely high energy deposition from positrons, electrons, photons and neutrons behind the positron target, and because a solenoid is required to focus the large emittance positron beam, the 1.3 GHz pre-accelerator has to use normal conducting structures up to energy of 400 MeV. There are many challenges in the design of the normal-conducting portion of the ILC positron injector system such as obtaining high positron yield with required emittance, achieving adequate cooling with the high RF and particle loss heating, and sustaining high accelerator gradients during millisecond-long pulses in a strong magnetic field. Considering issues of feasibility, reliability and cost savings for the ILC, the proposed design for the positron injector contains both standing-wave (SW) and traveling-wave (TW) L-band accelerator structures. A short version of the new type of the SW section is under fabrication and testing at SLAC[10].

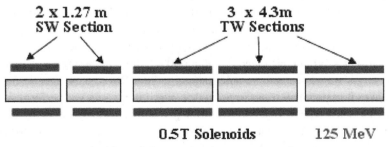

Figure 6. Layout of positron capturing region

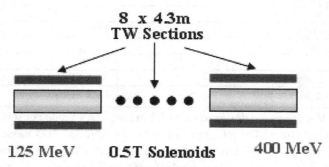

8 x 4.3m
TW Sections

125 MeV   0.5T Solenoids   400 MeV

Figure 7. Layout of positron Pre-Accelerator

As shown in figure 6, the capture region is composed of two 1.27 m SW accelerator sections at 15 MV/m accelerating gradient and three 4.3 m TW accelerator sections at 8.5 MV/m accelerating gradient in order to capture and accelerate the electron beam to 125 MeV. The positrons are then accelerated from 125 MeV to 400 MeV in a pre-accelerator region, which is composed of eight 4.3 m TW sections at 8.5 MV/m accelerating gradient as shown in figure 7. All accelerator sections are surrounded with 0.5 T solenoids.

The high gradient (15 MV/m) positron capture sections have been designed to be simple π mode 11 cells SW type of accelerator structures. The advantages are a more effective cooling system, higher shunt impedance with larger aperture (60 mm), lower RF pulse heating, apparent simplicity and cost savings. The mode and amplitude stability under various cooling conditions for this type of structure have been theoretically verified. Figure 8 shows a cutaway view of the SW structure and Table 2 gives the important RF parameters.

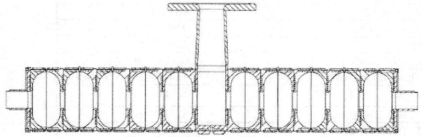

Figure 8. 11 Cell π mode SW structure

Table 2. Parameter of SW structure

Structure Type	Simple π Mode
Cell Number	11
Aperture 2a	60 mm
Q	29700
Shunt impedance r	34.3 MΩ/m
E_0 (8.6 MW input)	15.2 MV/m

All TW sections are designed to be 4.3 m long, $3\pi/4$ mode constant gradient accelerator structures. The RF group velocity of traditional $2\pi/3$ mode traveling wave structures is too high for our larger apertures (the radio of iris radius with wavelength a/λ~10%) to obtain a good RF efficiency. Therefore, to increase the "phase advance per cell" was used to optimize the RF efficiency for designing this type of large aperture TW structure. Compared with standing wave structures, the advantages are lower pulse heating, easy installation for long solenoids, no need to use circulators for RF reflection protection, apparent simplicity and cost saving. Figure 9 shows the shapes for three typical cells and Table 3 gives the important RF parameters.

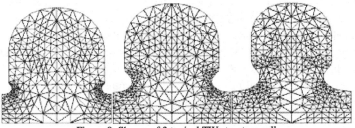

Figure 9. Shapes of 3 typical TW structure cell

Table 3. RF parameters of TW structure

Structure Type	TW $3\pi/4$ Mode
Cell Number	50
Aperture 2a	46 mm
Attenuation τ	0.98
Q	24842 - 21676
Group velocity Vg/c	0.62% – 0.14%
Shunt impedance r	48.60 – 39.45 MΩ/m
Filling time T_f	5.3 μs
Power Dissipation	8.2 kW/m
E_0 (8.6 MW input)	8.0 MV/m

4. The RDR undulator based positron source

As mentioned in previous section, the ILC RDR undulator based positron source is consisted of a helical undulator with K=0.92 and λu=1.15cm; a 2 meter diameter target wheel with a rim of 2cm wide, 1.4cm thick titanium alloy; an AMD with 5T field on target surface and decreased adiabatically down to 0.5T in 20cm followed by capturing RF system, etc. The drive beam is the 150GeV main electron beam. Simulation has shown that for 100m RDR undulator, the positron yield is ~1.28 for 100m long RDR undulator without photon collimator and ~0.7 for 100m long RDR undulator with photon collimator to enhance the positron beam polarization to ~60%. In order to achieve a yield of 1.5, one has to increase the length of undulator in both cases. For low polarization source, one will need about 117m long RDR undulator to achieve the 1.5 goal of positron yield. For 60% polarization, one will need ~215m long RDR undulator.

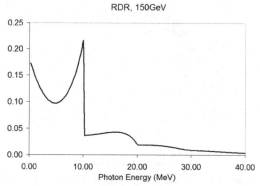

Figure 10. Photon number spectrum of RDR undulator with 150GeV drive.

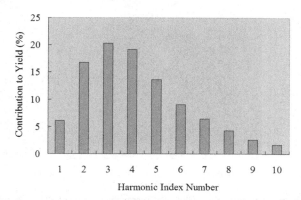

Figure 11. Yield contribution from different harmonics of RDR undulator.

Showing in figure 10 is photon number spectrum of RDR undulator with 150GeV drive beam. As shown in figure 10, the RDR baseline undulator has a 1st harmonic critical energy of ~10MeV. The contributions to the total number of photons from harmonics are ~52% for 1st harmonic, ~22% from 2nd harmonic, 11% from 3rd harmonic, 6% from the 4th harmonic, 3.6% from the 5th harmonic, 2% from the 6th harmonic, 1.2% from the 7th harmonic, 0.8% from the 8th harmonic. But due to the larger cross section of the positron production from higher energy photons, we found that these higher harmonics play dominant role in producing positrons in the target, even though their total number is smaller.

As shown in figure 11, the partition of yield contribution from harmonics for polarized positron source using undulator with K=0.92 λu=1.15cm, the 1st harmonic only contributes ~6%. The contribution to the captured positron beam is dominated by 2nd to 6th harmonic.

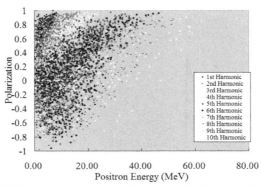

Figure 12. Initial positron polarization and energy distribution of RDR positron source exiting the target surface.

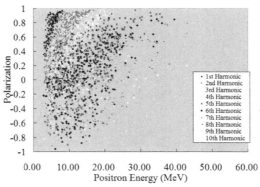

Figure 13. Initial positron polarization and energy distribution of RDR positron source captured positrons.

Showing in figure 12 and 13 are the initial polarization and energy distribution of positrons of the RDR positron source of all positrons and captured positrons as they exiting the target surface. Assuming the polarization will be preserved, then the polarization of captured positron beam will be ~25% if OMD is AMD or ~33% if OMD is FC.

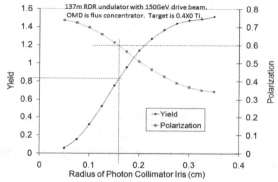

Figure 14. Yield and polarization of positron source with 137m RDR undulator using FC as OMD.

Showing in figure 14 is yield and polarization of positron beam of 137m RDR undulator with FC as function of photon collimator iris. As shown in this figure, 60% polarization can be achieved with a photon collimator of 1.6mm in radius of iris. The yield for 137m RDR undulator will be 0.82 when polarization is 60%. In order to have yield of 1.5 and polarization of 60%, the length of undulator has to be increased to about 256m.

5. The SB2009 undulator based positron source

As mentioned before, the undulator of positron source has been relocated to the end of electron main linacs in SB2009 proposal and also a quarter wave transformer has been chosen to be the OMD. As a result of these changes, the drive beam energy of undulator will be varying from 50GeV to 250 GeV for different running scenarios. As the quarter wave transformer has been used for OMD, in order to achieve the yield of 1.5, the length of undulator increased to 231m for nominal drive beam energy of 150GeV and thus the heat load in target increased.

As showing in figure 15, the photon number spectrum of the photon radiation from RDR undulator has a strong dependence on the drive beam energies. The number of photons stays the same for different drive beam energy, but the critical photon energy scales with γ^2. The pair production cross section strongly depends on the photon energy and thus the positron yield also strongly depends on the drive beam energy.

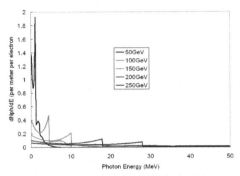

Figure 15. Photon number spectrums of RDR undulator with different drive beam energies.

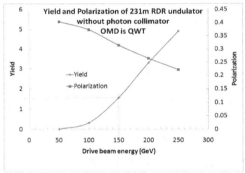

Figure 16. Yield and polarization of SB2009 undulator based positron source.

Showing in figure 16 are the yield and polarization of SB2009 undulator based positron source as a function of drive beam energy. As showing in this figure, the yield is 1.5 for the nominal drive beam energy, 150 GeV and the corresponding polarization is about 30%. The positron yield will drop significantly when drive beam energy goes below 100GeV. In order to maintain the luminosity of the machine at lower energy runs, 10Hz operation is introduced into the SB2009 proposal where 5 pulses will be at 150GeV for positron generation and 5 pulses will at the lower energy for collision. When the drive beam energy increased, the yield goes up while the polarization goes down. In order to maintain the polarization at about 30%, one can lower the magnetic field of undulator and thus lowered the K and increase the portion of photon from 1st harmonic and bring the polarization back. Study has shown that, for 250GeV drive beam and 231m long undulator with $\lambda u = 1.15$cm, one can run the undulator at K is about 0.45 to maintain the polarization of captured beam to be 30% while the yield is 1.5.

Table 4. Drive beam energy lost for SB2009 undulator based positron source

Drive beam energy	Energy lost per 100m	Energy lost for 1.5 yield
50GeV	~225MeV	N/A
100GeV	~900MeV	~9.9GeV
150GeV	~2GeV	~4.6GeV
200GeV	~3.6GeV	~3.7GeV
250GeV	~5.6GeV	~3.96GeV

Showing in table 4 are the drive beam energy lost for different drive beam energy passing through SB2009 undulator.

6. High K short period undulator

The current ILC RDR undulator is NbTi based superconducting helical undulator which has a K of 0.92 and period of 1.15cm. With this undulator, when the drive beam energy get lowered down to 100GeV, the positron yield will dropped down to ~0.3 from 1.5 (150 GeV drive beam, 137m long undulator, 0.4X0 Ti target and using flux concentrator). In order to achieve a reasonable yield at lower drive beam energy, it is required to push the period even shorter and keep the K at same level in the same time.

With Nb3Sn superconducting strand, it is possible to have a shorter period and high K in the meantime. The goal set by undulator design group is to reduce the period to ~9mm.

Showing in figure 17 is a scan of undulator parameter with a fixed drive beam energy of 100GeV. The assumptions used in the setup of this set of simulation are as follows:

- Length of undulator: 231m
- Drive beam energy: 100GeV
- Target: 0.4X0, Ti
- Photon Collimation: None
- Drift to target: 400m from end of undulator
- OMD: FC, 14cm long, ramping up from 0.5T to over 3T in 2cm and decrease adiabatically down to 0.5T in 12cm.

As shown in figure 17, the yield peaks around K=1.2 for all the different

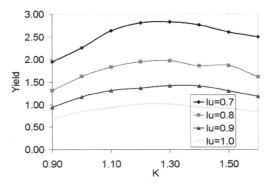

Figure 17. Yield of different high K short period undulator with 100GeV drive beam energy.

period length considered while increasing with the reducing of undulator period. For the nominal targeting parameter of new undulator, K=0.9 and λu=0.9cm, the yield of 231m long such undulator will have a positron yield of about 0.94 which is about twice the value for RDR undulator driven with 100GeV beam with the same length of undulator and capturing optics. From the results shown in figure 10, the advantage of short period and high K undulator is obvious. Some detail about high K short period undulator based positron source can found in [11]

7. Summary

An overview of the positron source for International Linear Collider is presented in this paper. In this overview, both RDR and SB2009 baseline positron source are covered together with the R&D status of critical components of undulator based positron source for ILC. Some of the parameters of positron source of both RDR and SB2009 baseline are also presented in this overview.

References

1. International Linear Collider Reference Design Report, Vol. 3, Aug, 2007
2. F. Zhou, etc. Start-to-end beam optics development and multi-particle tracking for the ILC undulator based positron source, SLAC-PUB-12239, Jan. 2007
3. F. Zhou, etc. Transport Optics Design and Multi-Particle Tracking for the ILC Positron source, pp: 3124-3126, Proceedings of PAC07, Albuquerque, New Mexico, USA
4. W. Liu, W. Gai, K.-J. Kim, *Systematic Study of Undulator Based ILC Positron Source: Production and Capture*, PAC07 – Proceedings, Albuquerque, New Mexico, USA, 2007
5. Owen Taylor, *Update on 4m module and generating higher fields*, 7th Positorn source collaboration meeting, Desy.
6. K. Floettmann, Positron Source Options for Linear Colliders, pp 69-73, Proceedings of EPAC 2004, Lucerne, Switzerland
7. Jeffrey Gronberg, private communication.
8. S. Antipov, W. Gai, W. Liu, L. Spentzouris. *Numerical Studies of International Linear Collider Positron Target and Optical matching device field effects on Beam.* Published in Appl. Phys. Vol. 102, 014910 (2007)
9. Ian Bailey, *Target eddy current experiment and modeling*, 7th Positorn source collaboration meeting, Desy.
10. J. W. Wang, etc, Positron Injector Accelerator and RF System for the ILC, SLAC-Pub-12412, 2007
11. W. Liu, W. Gai, Numerical Study on High K, Short Period Undulator for ILC Positron Source, ILC-NOTE-2011-058.

ON THE PARAMETERS OF INTERNATIONAL LINEAR COLLIDER POSITRON SOURCE

WANMING LIU[(1)], WEI GAI[(1)], SABINE RIEMANN[(2)], ANDRIY USHAKOV[(3)]

(1)Argonne National Lab, Argonne, IL, 60439, USA

(2) Deutsches Elektronen Synchrotron, Platanenallee 6, D-15738 Zeuthen, Germany

(3) II. Institute for Theoretical Physics, University of Hamburg, Luruper Chaussee 149, D-22761 Hamburg, Germany

The parameter set for the International Linear Collider (ILC) positron source given in the Technical Design Report (TDR) is more complicated than that presented in the ILC Reference Design Report (RDR). Studies to define and to optimize the parameters for different scenarios of center-of-mass energies have been performed at both Argonne National Lab (ANL) and Deutsches Elektronen Synchrotron (DESY)/Hamburg University. Results from both institutes agree well and are presented in this paper.

The ILC positron source is based on a helical undulator. As shown in Figure 1, the electron beam from the main linac will pass through a few hundred meter long undulator to create photon radiation. The electron beam will be sent to the Beam Delivery System while the photon beam will drift for about 400m to the conversion target to produce positrons through electromagnetic scattering. The properties of the positron beam coming out of the conversion target are related to the properties of the photon beam produced from the undulator. The spectra of the photon beam is determined by both the drive beam energy and the undulator parameters. Thus the parameters of the ILC positron source will change if the drive beam or the undulator parameters are changed.

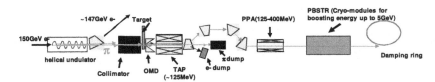

Figure 1. Conceptual layout of ILC positron source

Starting from SB2009 baseline of the ILC positron source, the site location of the undulator has been moved from the 150GeV position to the end of the electron main linac. As a result of this change, the parameters of the positron source are now tightly coupled with the center-of-mass energy scenarios of the proposed machine and a complete study to define these parameters is required. Two independent groups at ANL and DESY/Hamburg University performed many numerical studies and the results are presented in this paper.

1. The undulator parameters

In all previous design considerations, the 500GeV option was used to define the basic parameters for the undulator. The helical undulator for the ILC positron source has K=0.92 and λ_u=1.15cm [1]. The required yield of the positron source is 1.5 positrons per electron. Due to the recent progress in R&D of the flux concentrator (FC) [2], the FC has been established as baseline for the Optical Matching Device (OMD). With the FC, the capture efficiency is increased and thus the active length of undulator could be reduced from 231m (SB2009) down to 147m. Also, for E_{CM}=350GeV and 500GeV, the drive beam energy is higher than the nominal drive beam energy of 150GeV foreseen in the RDR. This allows to consider a lower B field of undulator keeping the length and the poitron yield unchanged. For center-of-mass energies smaller than 300GeV, it is quite challenging to produce the required number of positrons given the current undulator. Hence, a 10Hz operation scheme is adapted interlacing a 5Hz 150GeV drive beam for the positron production with the 5Hz luminosity beam. The corresponding parameters are listed in Table 1.

Table 1. Undulator parameters of ILC TDR Positron Source

Parameter	Ecm (GeV)				
	200	230	250	350	500
e+ production pulse rate (Hz)	5				
Drive beam energy (GeV)	150			175	250
Number of bunches per pulse	1312				
Number of e+ per bunch($\times 10^{10}$)	2				
1st harmonic energy (MeV)	10.1			16.2	42.8
Half opening angle of γ beam (μr)	3.4			2.9	2.0
Active length of undulator(m)	147				
Undulator K value	0.92			0.75	0.45
Undulator B field (T)	0.86			0.698	0.42
Undulator period length, λu (cm)	1.15				
Drive beam energy lost (GeV)	3.0			2.6	2.0
γ beam power (kW)	63.1			54.7	41.7

In SB2009, the effective length of the ndulator was 231m. This space is also allocated in the current TDR design to provide sufficient space for additional undulator modules required for the polarization upgrade.

2. Drive beam parameters

Starting from SB2009, the positron source has been relocated to the end of electron main linac and thus the source parameters will be coupled with the center-of-mass energy. When the electron beam energy is higher than the 150GeV nominal drive beam energy, the same undulator will give higher yield with lower polarization. In order to maintain the 30% baseline polarization, we lower the B field of undulator while keeping the undulator length constant as already shown in Table 1. Here in Table 2, we list the drive electron beam parameters for different center-of-mass energy scenarios.

Table 2. Drive electron beam parameters.

Parameter	E_{cm} (GeV)				
	200	**230**	**250**	**350**	**500**
e- linac pulse repitition rate(Hz)	10			5	
Number of bunches	1312				
Electron bunch population($\times 10^{10}$)	2				
Nominal 5Hz mode:					
Beam energy (GeV)	-			178	253
10Hz alternate pulse mode:					
Beam energy for e+ production (GeV)	150			-	
Beam energy for physics (lumi prod.) (GeV)	101	117	127	-	
Beam energy for e+ production (dumped) (GeV)	147			-	
Average beam power (e+ prod.) dumped (MW)	3.1			-	
Electron bunch separation (ns)	554				
Electron beam pulse length (μs)	727				
Electron pulse current (mA)	5.8				
Horizontal emittance (e+ prod.) (μm)				10	
Vertical emittance (e+ prod.)(nm)				35	

Effective undulator length (m)	147				
Effective undulator field (T)	0.86			0.698	0.42
undulator period length (cm)	1.15				
Electron energy loss in ndulator (e+ prod.) (GeV)	3.0			2.6	1.99
Electron energy loss in undulator (lumi prod.) (GeV)	1.3	1.8	2.1		
Rel. energy spread induced by und.(assumed initial 0.3%)	0.087	0.100	0.112	0.118	0.065
Total energy spread (assumed 0.3% initial)	0.312	0.316	0.320	0.322	0.307
Rel. energy spread induced by und.(assumed initial 0.2%)	0.092	0.112	0.117	0.116	0.082
Total energy spread (assumed 0.2% initial)	0.220	0.229	0.232	0.231	0.216
Rel. energy spread induced by und.(assumed initial 0.1%)	0.098	0.111	0.120	0.120	0.085
Total energy spread (assumed 0.1% initial)	0.140	0.149	0.156	0.156	0.132
Rel. energy spread induced by und.(assumed initial 0%)	0.098	0.113	0.123	0.122	0.084
Emittance growth (nm)	-0.4	-0.6	-0.7	-0.55	-0.31

As shown in Tables 1 and 2, for E_{CM}=375GeV and 500GeV, we lowered the undulator B field from 0.86T down to 0.698T and 0.42T. By doing this, the photon spectra has changed: the intensity of higher order harmonics is reduced and less photons are produced for a given undulator length. With this arrangement, the positron polarization can be increased to about 30% without adding a photon collimator. As reported in [3], the emittance of the electron beam will be reduced as the electron beam passes through the RDR undulator. The detailed explanation of this bahavior can be found in [3]. Another parameter worth noting in Table 2 is the undulator induced energy spread. The induced energy spread does not only depend on the beam energy but also on the initial energy spread of the electron beam. For higher beam energy more energy will be lost into photons and thus a higher energy spread is induced. For the same reason, a higher initial energy spread will induce less energy spread for the same beam energy.

3. Production target parameters

The ILC positron source production target is a 1m diameter wheel made of Ti-6%Al-4%V. The thickness of the target is 1.4cm corresponding to 0.4 radiation length of the target material. It will be spinning at a tangential speed of 100m/s. The distance between the target wheel and the end of the undulator is 400m. Since the photon spectra of a given undulator is determined by the drive beam energy, the other target parameters are coupled with the drive beam energy and are listed in Table 3.

Table 3. Production target parameters

Parameter	E_{cm} (GeV)				
	200	**230**	**250**	**350**	**500**
Captured positron yield (e+/e-)	1.5				
Positron polarization (%)	30.2			32.5	30.0
Positron pulse production rate (Hz)	5				
Electron beam energy (e+ prod.) (GeV)	150			178	253
Number of electron bunches	1312				
Electron bunch population($\times 10^{10}$)	2				
Photon energy (first harmonic) (MeV)	10.1			16.2	42.8
Photon opening angle (=1/γ) (μr)	3.4			2.9	2.0
Average photon power on target (kW)	91	100	107	55	42
Incident photon energy per bunch (J)	9.6			8.1	6.0
Energy deposition per bunch (e+ prod.) (J)	0.72			0.59	0.31
Relative energy deposition (%)	7%			7.20%	5%
Photon rms spot size on target (mm)	1.4			1.2	0.8
Peak energy density in target (J/cm^3)	232.5			295.3	304.3
Peak energy density in target (J/g)	51.7			65.6	67.5

As explained above, for E_{cm}<300GeV, the machine will be running at 10Hz mode with 5Hz 150GeV drive beam and 5Hz luminosity production beam. In Table 3, we listed only the numbers for the drive electron beam regarding the photon energy per bunch, the energy deposition per bunch, the relative energy deposition, the photon beam spot size on target (rms) and the peak energy density in the target. The peak energy density in the target is an accumulated effect which has taken into account the bunch separation and the target rotation speed. Since the B field of the RDR undulator has been lowered for

$E_{cm}>300GeV$, the incident photon energy per bunch has been reduced. But as the photon beam spot size on target is decreased also for increasing drive beam energy, the peak energy density in the target is increased for higher drive beam energy.

4. Capturing Optical Matching Device (OMD)

The optical matching device forseen in SB2009 was a quarter wave transformer which has a capture efficiency lower than 15%. With the recent progress in flux concentrator R&D[2], a flux concentrator with a peak field around 3.2T becomes feasible and beam dynamic simulation results shown in Figure 3

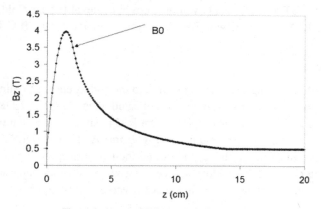

Figure 2. A typical FC on axis B field profile

demonstrate that the capture efficiency is high enough to qualify the FC being adapted as the TDR baseline OMD.

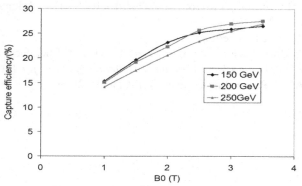

Figure 3. Capture efficiency of FC with different B0

As shown in Figure 2, a FC field profile is defined by its on-axis Bz field profile. We use this on-axis profile and paraxial approximation to obtain the full field map of FC used in the beam dynamic simulations. The on-axis B field profile is defined by the field on target (z=0), the position where the adiabatic decrease starts, the starting B field (B0) of adiabatic decreasing, the position where the adiabatic decreasing stops and the B field of the background. For the results presented in this paper, the B field on axis is fixed to 0.5T, the adiabatic decreasing starts at z=2cm and stops at z=14cm, and the background B field is 0.5T.

Figure 3 shows the capture efficiency of the FC depending on B0 for 3 different drive beam energies for the RDR undulator. The difference is small between B0=3.5T and B0=3T. Given the current achievable peak field in the FC designed by LLNL[2], a FC with B0=3T is recommended for the ILC TDR.

5. Summary

As a result of relocating the positron source to the end of electron main linac, the ILC TDR positron source is more complicated than the ILC RDR positron source. The results of studies to define the ILC positron source parameters are documented in this paper. Both groups from ANL and DESY/Hamburg University are involved in these studies and their results agree with each other. The parameters for TDR positron source polarization upgrade, upgrade to high energies and the luminosity upgrade are not covered in this paper.

References

1. ILC Reference Design Report, Volume3, Accelerator.
   http://ilcdoc.linearcollider.org/record/6321/files/ILC_RDR_Volume_3-Accelerator.pdf?version=4
2. J. Gronberg, these proceedings
3. W. Gai, M. Borland, K.-J. Kim, W. Liu, A. Xiao, J. Sheppard, *Emittance Evolution of the Drive Electron Beam in Helical Undulator for ILC Positron Source*, PAC09-Proceedings, Vancouver, Canada, 2009

POLARIZED POSITRONS FOR THE ILC — UPDATE ON SIMULATIONS

A. USHAKOV*, O. S. ADEYEMI and G. MOORTGAT-PICK

*II. Institute for Theoretical Physics, University of Hamburg,
Luruper Chaussee 149, D-22761 Hamburg, Germany*
* E-mail: andriy.ushakov@desy.de

F. STAUFENBIEL and S. RIEMANN

*DESY Standort Zeuthen
Platanenallee 6, 15738 Zeuthen, Germany*

To achieve the extremely high luminosity for colliding electron-positron beams at the future International Linear Collider[1] (ILC) an undulator-based source with about 230 meters helical undulator and a thin titanium-alloy target rim rotated with tangential velocity of about 100 meters per second are foreseen. The very high density of heat deposited in the target has to be analyzed carefully. The energy deposited by the photon beam in the target has been calculated in FLUKA. The resulting stress in the target material after one bunch train has been simulated in ANSYS.

Keywords: ILC; positron source; thermal stress.

1. Introduction

The positron-production target for the ILC positron source is driven by a photon beam generated in an helical undulator placed at the end of main electron linac.[2] The undulator length is chosen to provide the required positron yield. The source is designed to deliver 50% overhead of positrons. Therefore, the positron yield has to be 1.5 positrons per electron passing the undulator. The required active length of the undulator is about 75 meters for the nominal electron energy of 250 GeV, the undulator K-value has been chosen to be 0.92, the undulator period is 11.5 mm and a quarter-wave transformer is used as optical matching device (OMD). The photon first harmonic energy cutoff is 28 MeV, the average energy of photons is about 29 MeV and the average photon beam power is about 180 kW in a train of 2625 bunches with a frequency of 5 Hz. Although only relatively small fraction of total photon beam energy deposited in the target (about 5%), the peak energy density deposited in target is high due to the small opening angle of the synchrotron radiation in the helical undulator resulting in a small photon spot size on the target. For example, for 500 meters space between the undulator and target, the average radius of the photon beam is approximately 2 mm and the peak energy density could achieve 120 J/g in the 0.4 radiation length thick titanium-alloy target rotated with 100 m/s tangential velocity.

There is no experimental data indicating the upper limit of the peak energy density deposited by photons in the titanium alloy material with 90% of titanium, 6% of aluminium and 4% of vanadium. The analysis of the electron beam induced damage to the SLC positron target[3] and the simulations of thermal shock[4] show that the energy deposition limit is about 30 J/g for tungsten with 25% of rhenium target irradiated by 33 GeV electrons and a general criteria of failure due to an equivalent (von-Mises) stress of 50% of tensile strength may apply to this target material.[4]

The thermal structural modeling of a rotated titanium target irradiated by helical undulator photons has been performed for the NLC by W. Stein and J. Sheppard.[5] They recommend to consider as "safe" thermal stresses below one third to one half of the yield stress.

In this paper, the thermal stress in the ILC positron source target has been estimated for the SB2009 set of parameters.[2]

2. Energy Deposition and Temperature in Target. Static Model of Material Response

The energy transfer from the photon beam into temperature of target material and the structural deformation and mechanical stress coupled with this temperature rise due to complexity of these time-dependent, cross-coupled and nonlinear processes cannot be treated with the highest level of details.[6] Therefore, the choice of simulation tools and reasonable approximations and simplifications plays an important role.

The energy deposition in the target has been calculated in FLUKA.[7] An amount of energy is counted as deposited if after collisions the primary or secondary particles have energies lower than the energy cut-offs. The FLUKA default cut-offs were used: 1.511 MeV – for electrons and positrons and 333 keV – for photons.

Figure 1 shows the "original" FLUKA data distribution (i.e. without any scaling factors) of the energy deposited close to the back side of the target. The energy is given in units of GeV per cubic centimeter and per

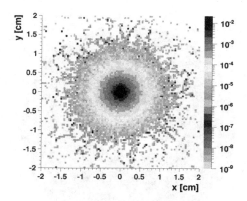

Fig. 1.   Distribution (dependence in x and y) of the energy [GeV/(ph cm^3)] deposited close to the backside of Ti6Al4V target calculated in FLUKA.

impinging on the target photon.

The temperature rise δT in the target for given a energy deposition E_{dep} has been calculated according to the following equation

$$\delta T = \frac{E_{dep} N_{e^-} Y_{ph} L_{und} N_b}{\rho c_p},$$

where N_{e^-} is the number of electrons per bunch (2×10^{10}), Y_{ph} is the photon yield (1.94 photons per electron and per 1 meter of undulator), L_{und} is the length of undulator (70 meters), N_b is the number of bunches crossing the same volume/bin, ρ is the target density (4.49 g/cm^3) and c_p is the specific heat capacity (0.523 J/(g K)).

The temperature data in a 1.48 cm thick cylindrical titanium target after the first 100 bunches has been imported into ANSYS.[8] The temperature distribution on the back side of the target is shown in Figure 2. The maximal increase of temperature per bunch is about 2.2 K.

As a first step, a statical ANSYS model of the target material response to the heat load (see, Figures 1 and 2) has been applied. The total deformation and equivalent von-Mises stress are shown in Figs. 3 and 4. The maximum of equivalent stress is about 100 MPa on the back side of the target in the circular area around the photon beam axis with a radius of approx. 2 mm. This stress is about 12% of the tensile yield strength for titanium alloy (the properties of Ti6Al4V alloy, grade 5 can be found, for example, in Ref.[9]).

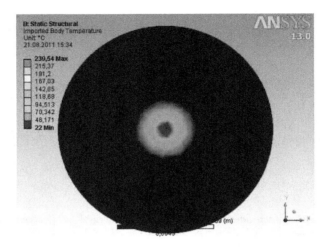

Fig. 2. Temperature profile on target back side after 100 bunches.

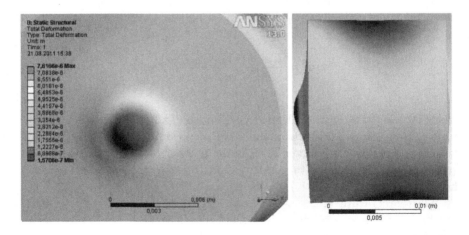

Fig. 3. Total deformation of the target after 100 bunches (back view – left, side view – right).

3. Evolution of Thermal Stress in Time

To simulate the time evolution of thermal stress in the positron source target, the target movement has been analyzed more accurately and ANSYS transient (explicit) model of deformation and stress has been used.

We consider the tangential velocity of the target rim (1 meter in di-

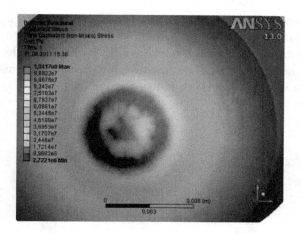

Fig. 4. Von-Mises stress after 100 bunches (ANSYS static structural model).

ameter) of 100 meters per second as velocity in y direction in a Cartesian system. The energy deposited after one pulse (1312 bunches with 554 ns bunch separation) as function of y coordinate is shown in Fig. 5. This Figure shows also the energy deposited by a single bunch and the corresponding temperature rise. Both profiles on Fig. 5 are plotted for highest energy deposition: in the z-direction – close to the target back side and in the middle of bunch(es) – in x-direction. The bunch overlapping factor is defined as the ratio of the maximal deposited energy after a complete bunch train with respect to the maximum after just one bunch. This factor for the nominal SB2009 source parameters is about 59.

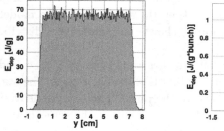 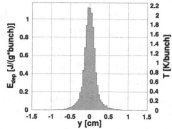

Fig. 5. Energy deposition in a target rotated with 100 m/s: left – after one pulse, right – after one bunch.

Figure 6 shows the temperature and equivalent stress in the "rotated" target. The target has been cut in the middle plane in order to show the distributions inside the target. The static ANSYS model for the equivalent stress after one bunch train does not take into account thermal diffusion and thus overestimates the stress induced in the target.

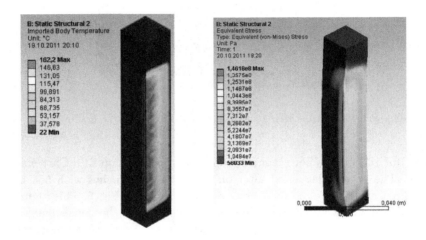

Fig. 6.   Temperature distribution and induced equivalent stress in the rotated target.

To reduce the effect of thermal diffusion on the stress and to study the time-dependent dynamic effects, another model has been used. In this model the temperature distribution after one single bunch has been scaled with the above-mentioned bunch overlapping factor. The cylindrical geometry of the target has been chosen to keep the symmetry of the model and to reduce the computing time. Figure 7 shows the temperature distribution after "59 bunches".

The total deformation after 59 bunches is plotted in Fig. 8 and the evolution in time of maximal deformation is shown in Fig. 9. The starting time (0 sec.) corresponds to the end of the pulse. The reflections from the target surfaces and interference of the waves result in the series of maxima at the level about 25% of the initial deformation.

The deformation transverse to the beam axis (radial deformation) contributes only minor (about one third) to the total deformation. The time

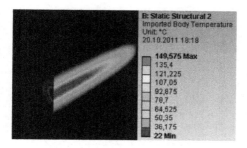

Fig. 7. Temperature distribution after 59 bunches.

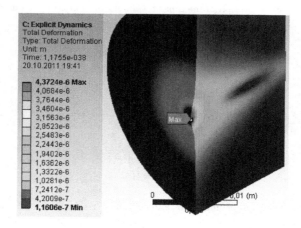

Fig. 8. Distribution of total deformation after 59 bunches.

dependence of the dominating longitudinal (z-component) velocity is presented in Fig. 10 showing the positive velocity directed out of the target and negative velocity. Figure 11 shows a snapshot of the time evolution for the v_z-distribution after one pulse and with 0.1 μs delay at the moment when the negative velocity has reached the maximum. The y-component of deformation and velocity are also shown in Figs. 12 and 13. Because of geometry and beam symmetry, the deformation, maximal and minimal v_y-dependencies on time are symmetrical (mirrored) too. It has to be noted that the transient effects during the pulse were not considered and thus the all velocities are starting from zero level at the end of the beam pulse.

The time evolution of the maximal equivalent (von-Mises) stress in the target is plotted in Fig. 14. The stress distributions after one beam pulse

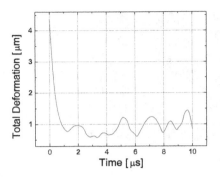

Fig. 9. Time evolution of maximal total deformation after 59 bunches.

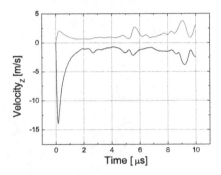

Fig. 10. Time evolution of maximal and minimal z-component of velocity after 59 bunches.

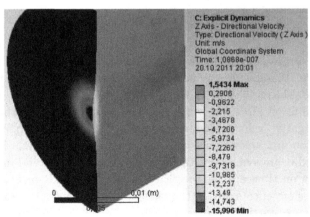

Fig. 11. Velocity along z-axis after 59 bunches and 0.1 μs delay.

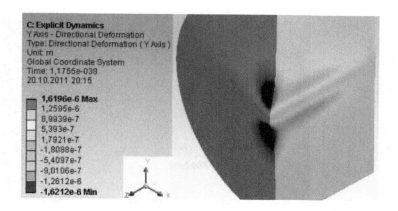

Fig. 12. Direction (y-axis) deformation after 59 bunches.

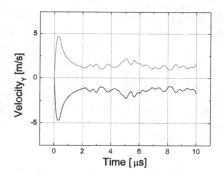

Fig. 13. Time evolution of maximal and minimal y-component of velocity after 59 bunches.

and additional 0.1 μs delay are shown in Fig. 15. The peak stress value is about 160 MPa which corresponds to 18% of tensile yield strength. This level of stress can be considered as acceptable.

4. Summary

The energy deposition in the ILC positron source target has been simulated in FLUKA for the SB2009 set of parameters. The peak energy density in the rotated titanium-alloy target is about 120 J/g for the conservative choice of a magnetic focusing device (quarter-wave transformer) and 250 GeV electron beam energy. The different simplified (static and transient) ANSYS models have been used to estimate the thermal stress induced by fast tem-

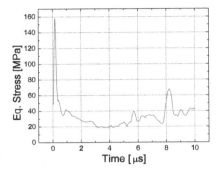

Fig. 14. Time evolution of maximal equivalent stress after 59 bunches.

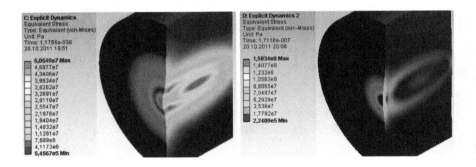

Fig. 15. Equivalent stress after 59 bunches (left) and additional 0.1 μs delay (right).

perature rise and thermal expansion of the target. The peak stress is about 160 MPa. It is less then 20% of tensile yield strength. Such stress will not damage the target.

5. Outlook

In the future, also the cooling of the target has to be added in model. The procedure used so far, in which the deposited energy is converted into temperature, has to be eliminated and the direct import of the heat source into ANSYS can additionally improve the accuracy of stress estimations. The thermal and structural effects in the target have to be also simulated taking into account the time structure of the bunch train.

Acknowledgments

We would like to thank the organizers and the host of POSIPOL 2011 for this fruitful and encouraging workshop and for hospitality.

Work supported by the German Federal Ministry of Education and Research, Joint Project R&D Accelerator "Spin Management", contract No. 05H10GUE.

References

1. N. Phinney, N. Toge and N. Walker (eds.), *International Linear Collider Reference Design Report: Volume 3: Accelelator* August 2007.
2. *ILC SB2009 Proposal Document* (December, 2009). http://ilc-edmsdirect.desy.de/ilc-edmsdirect/file.jsp?edmsid=D00000000900425.
3. V. K. Bharadwaj, Y. K. Batygin, J. C. Sheppard, D. C. Schultz, S. Bodenstein, J. Gallegos, R. Gonzáles, J. Ledbetter, M. López, R. Romero, T. Romero, R. Rutherford and S. Maloy, *Proc. of PAC 2001* , 2123 (2001).
4. W. Stein, A. Sunwoo, V. K. Bharadwaj, D. C. Schultz and J. C. Sheppard, *Proc. of PAC 2001* , 2111 (2001).
5. W. Stein and J. C. Sheppard, *NLC polarized positron photon beam target thermal structural modeling*, Tech. Rep. SLAC-TN-03-045, LCC-0087, UCRL-ID-148940, SLAC (Stanford, CA, 2002).
6. J. M. Zazula, *Proc. of the 2nd Workshop on Simulating Accelerator Radiation Environments, CERN, Geneva, Switzerland, 9 - 11 Oct 1995* , 26 (1995).
7. A. Fassò, A. Ferrari, J. Ranft and P. R. Sala, *FLUKA: a multi-particle transport code*, Tech. Rep. CERN-2005-10, INFN TC_05/11, SLAC-R-773, CERN (Geneva, 2005).
8. *ANSYS*. http://www.ansys.com.
9. Aerospace Specification Metals, Inc., Titanium Ti6Al4V (Grade 5), Annealed http://asm.matweb.com/search/SpecificMaterial.asp?bassnum=MTP641.

WHY K-FACTOR IN ILC UNDULATOR SHOULD BE SMALL

ALEXANDER A. MIKHAILICHENKO,

Cornell University, LEPP, Ithaca, New York

Abstract. We are analyzing the ILC positron source for the best polarization and efficiency. We represent the arguments why K-factor ($K = eH\lambda_u / 2\pi mc^2$) should be ≤ 0.45. Lower K-factor allows reduction the total power of radiation and dominance the first harmonic in radiation. As the first harmonic can be well collimated, hence more monochromatic, the energy separation of the secondary positrons/electrons at high edge of spectrum can be done easily. As a result ~70% polarization looks feasible for ~170 m undulator at 150-500 *GeV* and K factor ≤0.4.

Overview

In ILC positron source the undulator scheme [1] is appointed as a baseline. Undulator radiation (UR) generated by particles with energy E~150-250 *GeV* serves as a source of circularly polarized gammas with energy $\hbar\omega$~10-20 *MeV*. So called K-factor ($K = eH\lambda_u / 2\pi mc^2 = \beta_\perp\gamma$ - where $\beta_\perp$ stands for the transverse velocity of electron normalized by the speed of light, $\gamma = E / mc^2$, λ_u is a period of magnetic field in the undulator)–is the mostly important characteristic, which defines the properties of UR. These properties include the harmonic content, i.e. its spectrum and polarization. So the momentum associated with the transverse oscillation of the particle comes to be $p_\perp = mc\beta_\perp\gamma = mcK$. The total energy radiated by a particle in the undulator with total number of periods M is proportional to the K^2; more exactly to the square of magnetic field B averaged over period [2]-[4]

$$\Delta\varepsilon_{tot} = \tfrac{2}{3}r_e^2\overline{B^2}\gamma^2 M\lambda_u = \tfrac{8\pi^2}{3}e^2 MK^2\gamma^2 / \lambda_u \qquad (1)$$

One can see from (1) that the total intensity of radiation is not a function of undulator period. Dependence of intensity on magnetic field squared limits the maximal energy, which any particle might have after passage the region with the field (Pomeranchuk theorem [5]).

The energy distribution of undulator radiation emitted by a single particle in an undulator with the length $L_u = M\lambda_u$ (M –is the number of periods), during the time duration $\Delta t = 2\pi M / \Omega$ is defined by the expression [2]-[7]

$$
\frac{d\varepsilon_n}{do} = \frac{dI_n}{do} \frac{2\pi M}{\Omega}
$$

$$
= \frac{e^2 \omega_n^3 M}{cn\Omega^2} [\beta_\perp^2 J_n'^2 \left(\frac{n\beta_\perp \sin\vartheta}{1 - \beta_\parallel \cos\vartheta} \right) +
$$

$$
\frac{(\cos\vartheta - \beta_\parallel)^2}{\sin^2\vartheta} J_n^2 \left(\frac{n\beta_\perp \sin\vartheta}{1 - \beta_\parallel \cos\vartheta} \right)]
$$

(2)

where $\beta_\perp = v_\perp / c = K / \gamma$, $v_\perp$ is the transverse velocity. In the dipole approximation, $K<1$, $\beta_\parallel = \sqrt{\beta^2 - \beta_\perp^2} \cong \beta$ and in the ultra-relativistic case $\gamma \gg 1$, mainly the first harmonic radiated, $n=1$.

$$
\frac{d\varepsilon_1}{do} = \frac{e^2 \Omega M}{c} \cdot F(\vartheta)
$$

$$
\cong \frac{e^2 \Omega M}{c} \cdot \frac{\beta_\perp^2}{8\pi(1 - \beta_\parallel \cos\vartheta)^3} \left[1 + \frac{(\cos\vartheta - \beta_\parallel)^2}{(1 - \beta_\parallel \cos\vartheta)^2} \right]
$$

(3)

where the function $F(\vartheta)$ introduced accordingly. Indeed, for total radiation at all harmonics, expression (2) should be summarized over all indices n [6]

$$
\frac{dI}{do} = \sum_{n=1}^{\infty} \frac{dI_n}{do}
$$

$$
= \frac{e^2 \Omega^2 \beta^2}{32c} \cdot \left[\frac{4 + 3\beta^2 \sin^2\vartheta}{(1 - \beta^2 \sin^2\vartheta)^{5/2}} + \frac{\cos^2\vartheta \cdot (4 + \beta^2 \sin^2\vartheta)}{(1 - \beta^2 \sin^2\vartheta)^{7/2}} \right]
$$

(4)

So the number of photons can be obtained just by division of intensity of radiation within some spectral region by the energy of the photon within this energy. The photon energy and its polarization for a given K factor depend on the observation angle ϑ measured from the longitudinal axis to the direction towards an observer

$$\hbar\omega_n = \frac{n\hbar\Omega}{1-\beta_\parallel \cos\vartheta} \; , \tag{5}$$

where $n=1,2,3...$, numerates the harmonics of frequency $\Omega = \beta_\parallel c / \lambda_u$, $\bar{\beta} = \bar{v}/c \cong \beta_\parallel$, $\bar{v}$ is a particle's average longitudinal velocity in the undulator. Basically this formula represents the Doppler shift of radiation while the particle oscillates with frequency Ω.

$$\frac{dN_n}{do} = \frac{d\varepsilon_n}{\hbar\omega_n do}$$

$$= \frac{dI_n}{do} \frac{2\pi M}{\hbar\omega_n \Omega} \tag{6}$$

$$= \frac{e^2}{\hbar c} \frac{\omega_n^3 M (1-\beta_\parallel \cos\vartheta)}{n^2 \Omega^3} \left[\beta_\perp^2 J_n'^2 \left(\frac{n\beta_\perp \sin\vartheta}{1-\beta_\parallel \cos\vartheta} \right) \right.$$

$$\left. + \frac{(\cos\vartheta - \beta_\parallel)^2}{\sin^2\vartheta} J_n^2 \left(\frac{n\beta_\perp \sin\vartheta}{1-\beta_\parallel \cos\vartheta} \right) \right]$$

By introduction of functions [3]

$$F_n(K,\vartheta) = F_n^+ + F_n^- \; ,$$

$$F_n^\pm(K,\vartheta) = \frac{1}{2}\left(J_n'(n\kappa) \pm \frac{\cos\vartheta - \beta_\parallel}{1-\beta_\parallel \cos\vartheta} \cdot \frac{1}{\kappa} J_n(n\kappa) \right)^2 , \tag{7}$$

$$F_n(K,\vartheta) = J_n'^2(n\kappa) + \left(\frac{\cos\vartheta - \beta_\parallel}{\beta_\perp \sin\vartheta} \right)^2 J_n^2(n\kappa) , \tag{8}$$

where $\kappa = \dfrac{\beta_\perp \sin\vartheta}{1-\beta_\parallel \cos\vartheta}$, intensity and polarization can be expressed as [3], [6]

$$\frac{dI_n}{do} = \frac{dI_n^+}{do} + \frac{dI_n^-}{do} \tag{9}$$

$$\frac{dI_n^\pm}{do} = I_{tot} \cdot \left[\frac{3}{4\pi\gamma^4} \frac{n^2 F_n^\pm(K,\gamma)}{(1-\beta_\parallel \cos\vartheta)^3} \right] \equiv I_{tot} \cdot P_n^\pm(K,\vartheta,\gamma) \tag{10}$$

$$\xi_{2n} = \frac{F_n^+ - F_n^-}{F_n} \equiv \frac{F_n^+ - F_n^-}{F_n^+ + F_n^-} = \frac{\dfrac{\partial I_n^+}{\partial\vartheta} - \dfrac{\partial I_n^-}{\partial\vartheta}}{\dfrac{\partial I_n^+}{\partial\vartheta} + \dfrac{\partial I_n^-}{\partial\vartheta}}$$

$$\tag{11}$$

$$= 2\left(\frac{\cos\vartheta - \beta_\parallel}{\beta_\perp \sin\vartheta} \right) \times \frac{J_n'(n\kappa)J_n(n\kappa)}{F_n(K,\vartheta)}$$

In formula (11) the angular dependence of intensity is represented as multiplication of two factors: the intensity of radiation I_{tot} and some function of angles, energy and K-factor $P_n^\pm(K,\vartheta,\gamma)$. As the integral over solid angle should coincide with the total intensity

$$\int_{4\pi} \frac{dI_n^\pm}{do} do = I_{tot} \cdot \int_{4\pi} P_n^\pm(K,\vartheta,\gamma) do = I_{tot}, \tag{12}$$

so the integral $\int_{4\pi} P_n^\pm(K,\vartheta,\gamma) do = 1$, hence $P_n^\pm(K,\vartheta,\gamma)$ can be treated as a *probability* of radiation of spectral component with certain helicity in direction defined by the angle ϑ. The spectral distribution one can obtain from the distribution over the angles, as the energy of quanta registered by an observer under angle ϑ is strictly connected with this angle through the formula (5). From the formula (5) it is clear also, that narrowing the angle ϑ by collimation– narrows the energy spread in the photon beam. This peculiarity of presentation the radiation as a sum of *probabilities* at different harmonics used in a computer code KONN [8], see below.

Operation with low K factor which we advocated for a long time reduces the content of higher harmonics as the first harmonic power reaches the 50% of all radiated power at $K=0.7$ only; we believe that $K<0.4$ is the best choice, see Fig.1.

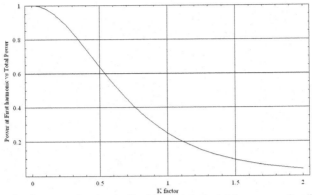

Figure1. The ratio of power radiated at the first harmonic to the total power as a function of K factor.

One peculiarity of radiation in a helical undulator is that the intensity of radiation in a straight forward direction presented by the first harmonic only; the other ones have zero intensity in this (forward) direction. So basically there is some possibility to keep the K factor increased, if some collimator in front of the target is present.

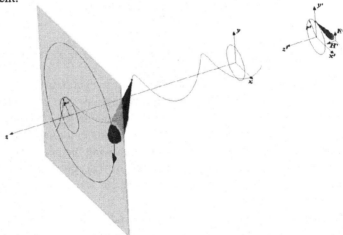

Figure2. Orbital angular momentum appears as the particle radiates from the off-centered trajectory (helix) [10].

One should remember, that the photons radiated at harmonic #2 and higher, are carrying the orbital momentum, so the creation of positron pairs is going differently; this fact was missed so far. Orbital angular momentum in radiation appears as the particle radiates from the helical trajectory; simple physical explanation of this phenomenon is given in [10] (see Fig.2). So for reduction of possible negative sequences of the fact that created electron-positron pair should carry the orbital momentum it is better to have the content of second (and higher) harmonic as low as possible.

Some Technical Details [9]

Axial magnetic field generated by a pair of thin helical strips-like conductors caring opposite current values $\pm I$, wounded on cylindrical surface (see Fig.3) can be found from the following expression [11]

$$H_\phi(\rho,\phi,z) = -\frac{I}{\pi\rho}\cdot\left(\frac{2\pi a}{\lambda_u}\right)\cdot\frac{\sin(\alpha)}{\alpha}\cdot$$

(13)

$$\cos\left(\phi-\phi_0-\frac{2\pi z}{\lambda_u}\right)\times I_1\left(\frac{2\pi\rho}{\lambda_u}\right)\left[K_0\left(\frac{2\pi a}{\lambda_u}\right)+K_2\left(\frac{2\pi a}{\lambda_u}\right)\right]$$

(SI units) where a stands for the radius of the windings, ϕ_0 is the local angle between the center of the strip and the axis x, ρ -is transverse radial coordinate, $K_0(2\pi a/\lambda_u)$, $K_2(2\pi a/\lambda_u)$, $I_1(2\pi a/\lambda_u)$ are the Bessel functions of the second kind, 2α represents the angle under which the strip is visible from the central axis. One would like to have period of undulator as small as possible $\lambda_u \to 0$, however $K_m(x) \approx e^{-x}\sqrt{\frac{\pi}{2x}}[1-\frac{1}{8x}+...]$ for $x\gg1$. In typical case, the diameter of windings is not more, than the period of windings, so the ratio $2\pi a/\lambda_u \equiv \pi(2a)/\lambda_u \cong \pi/2$, so $K_0(\pi/2)+K_2(\pi/2) \cong 0.71$.

Figure 3. B-helical coils (colored red and blue) that can generate a helical field. The beam is running inside the vacuum chamber colored yellow

Using expansion of Bessel functions I_1, (see for example [13]) $I_1(x) = \frac{x}{2} + \frac{x^3}{2^2 4} + ...$ one can obtain dependence of magnetic field on the transverse coordinate, $x = 2\pi\rho / \lambda_u \cong \rho / \lambdabar_u$

$$H_\phi = -\frac{0.71}{2}\frac{I}{\lambdabar_u} \cdot \cos\left(\phi - \phi_0 - \frac{2\pi z}{\lambda_u}\right) \times I_1\left(\frac{2\pi\rho}{\lambda_u}\right) \Big/ \left(\frac{2\pi\rho}{\lambda_u}\right) \cong$$

$$\cong -\frac{0.71}{4}\frac{I}{\lambdabar_u} \cdot \cos\left(\phi - \phi_0 - \frac{2\pi z}{\lambda_u}\right) \times \left[1 + \frac{1}{8}\left(\frac{\rho}{\lambdabar_u}\right)^2 + \frac{1}{768}\left(\frac{\rho}{\lambdabar_u}\right)^4 + ...\right] \quad (14)$$

So one can see that the period of undulator is restricted from the lower side by achievable current in a windings. For the undulator with period $\lambda_u \sim 1cm$ the typical K value is limited by a factor of one, $K \leq 1$. From the other hand- lower the K factor -larger the aperture in the undulator can be.

Polarization and the Energy Separation

The most valuable feature of the conversion using the undulator radiation is the possibility of generation of polarized positrons and electrons. Polarization of created positron as function of its energy E_+ can be expressed as [14]

$$\vec{\varsigma} = \xi_2 \cdot [f(E_+ / E_\gamma) \cdot \vec{n}_\parallel + g(E_+ / E_\gamma) \cdot \vec{n}_\perp] = \vec{\varsigma}_\parallel + \vec{\varsigma}_\perp, \quad (15)$$

where the functions f and g describe the longitudinal and transverse polarizations shown in Fig.4, $\vec{n}_\parallel$ – is the unit vector directed along initial direction of the gamma radiation, $\vec{n}_\perp$ –is the unit vector normal to it.

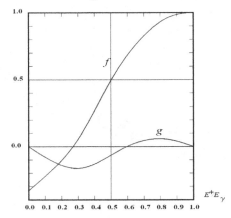

Figure 4. Polarization as function of the fractional energy [14].

One can see from Fig.4 that polarization of the secondary particles is the result of the product of two factors: the polarization of the photon and the function defined by the details of electron-positron pair creation. Expression (15) should be accompanied by the differential cross section of the pair creation [15]. So the polarization of the photon beam is a maximal possible value of polarization of the secondary particles. Namely this first factor one can enhance by collimation of the photon beam as the

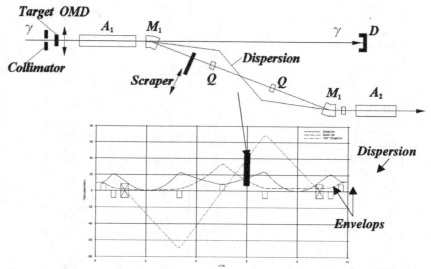

Figure 5. A scheme for the energy separation [16]. $A_{1,2}$ –accelerator structure, Q –is the focusing quadrupole M_1 –is the bending magnet, OMD stands for the Optical Matching Device.

It is interesting to underline, that in case if the first harmonic dominates in radiation, the role of collimator in front of the target is less important: if the energy selection tuned so that the only energetic particles are coming through, then automatically this procedure will select the particles with highest degree of photon polarization as these photons will generate the mostly energetic positrons. This is because the specifics of formula (11): higher polarization-higher the energy of quanta is.

Calculations with Konn

To simplify the process of optimization of positron conversion system we developed the interactive start-to end simulation code KONN, see [8]. This code realizes the ideas about presentation of radiation formulas (10) and (11) in terms of probability and uses the Monte-Carlo algorithm describing the processes of radiation and further creation of positron in a target. Results of some optimizations with KONN represented in Table 1. Lithium lens is used in collection optics as focusing element.

Table 1. Efficiency and polarization achievable with undulator scheme (KONN).

Beam energy, GeV	150	250	350	500
Length of undulator, m	180	200	200	200
K factor	**0.45**	**0.44**	**0.35**	**0.27**
Period of undulator, cm	1.0	1.0	1.0	1.0
Distance to the target, m	150	150	150	150
Radius of collimator, cm	0.049	0.03	0.02	0.02
Emittance, $cm \cdot rad$	1e-9	1e-9	1e-9	1e-9
Bunch length, cm	0.05	0.05	0.05	0.05
Beta-function, m	400	400	400	400
Thickness of the target/X_0	0.57	0.6	0.65	0.65
Distance to the length, cm	0.5	0.5	0.5	0.5
Radius of the length, cm	0.7	0.7	0.7	0.7
Length of the length, cm	0.5	0.5	0.5	0.5
Gradient, MG/cm	0.065	0.065	0.08	0.1
Wavelength of RF, cm	23.06	23.06	23.06	23.06
Phase shift of crest, rad	-0.29	-0.29	-0.29	-0.29
Distance to RF str., cm	2.0	2.0	2.0	2.0
Radius of collimator[†], cm	2.0	2.0	2.0	2.0
Length of RF str., cm	500	500	500	500
Gradient, MeV/cm	0.1	0.1	0.1	0.1
Longitudinal field, MG	0.045	0.045	0.045	0.045
Inner rad. of irises, cm	3.0	3.0	3.0	3.0
Acceptance, $MeV \cdot cm$	5.0	5.0	5.0	5.0
Energy filter, $E > -MeV$	54	74	92	126
Energy filter, $E < -MeV$	110	222	222	250
Efficiency, e^+/e^-	**1.5**	**1.5**	**1.5**	**1.5**
Polarization, $\%$	**69**	**78**	**78**	**73**

† Collimator at the entrance of RF structure

Summary

Practical value of K factor for the undulator with 1-cm period is about unity, which is enough for successful operation of conversion system.

Selection of energy with dispersion optics and the scraper allows enhancement of polarization and operation with increased K-factor up to $K \sim 0.4$. Lowering the K factor allows larger aperture in undulator.

As the radiation of electron in a back- scattered radiation from a laser, can be described in a same way as the radiation in an undulator [17] while the energy of secondary photon is much lower than the energy of electron, so the recommendation for lowering K factor automatically fulfilled here due to limitation of power achievable in a laser system.

Operation with low K-factor in E-166 experiment ($K\sim0.17$) together with selection the energy of the secondary positrons by the spectrometer delivered the polarization measured of the order $\varsigma_{\parallel} \cong 85\%$ [18].

References

1. J.A.Clarke *et al.*, *"The Design of the Positron Source for the International Linear Collider"*, EPAC08-WEOBG03, Jun 25, 2008. 3pp., Conf. Proc C08-06-23.3 (2008).
2. A.A.Sokolov, I.M.Ternov *et al.*, Izvestia Vuzov Fizika, N 5, 43 (1968); Zs. F. Phys. **211**, 1 (1968).
3. D.F.Alferov, Yu.A.Bashmakov, E.G.Bessonov, *"Undulator Radiation"*, Sov. Phys. Tech. Phys., 1974, v.18, No 10, p.1336.
4. B.M.Kincaid, *A Short-Period Helical Wiggler as an Improved Source of Synchrotron Radiation"*, Journal of Applied Physics, Vol.48, No 7, July 1977.
5. L.D.Landau, E.M.Lifshits, *"The Classical Theory of Fields"*, Pergamon Press.
6. A.A.Sokolov, I.M.Ternov," *Relativistic Electron"*, Moscow, Nauka (UDK 539.12) 1974.
7. E.Bessonov, A.Mikhailichenko, *"Coherent Radiation in Insertion Devices-II"*, CBN 11-5, August 23, 2011, available at http://www.lepp.cornell.edu/public/CBN/2011/CBN11-5/CBN%2011-5.pdf
8. A.Mikhailichenko. *"Update of Positron Production Code KONN"*, LCWS Chicago, Nov. 15-20, 2008.
   http://ilcagenda.linearcollider.org/materialDisplay.py?contribId=502&sessionId=10&materialId=slides&confId=2628
9. A. Mikhailichenko," *Pulsed Helical Undulator for Test at SLAC the Polarized Positron Production Scheme. Basic description"*, CBN 02-10, September 16, 2002;
   http://www.lns.cornell.edu/public/CBN/2002/CBN02-10/CBN02-101.pdf
10. A.Afanasev, A. Mikhailichenko, *"On Generation of Photons Carrying Orbital Angular Momentum in the Helical Undulator"*, Sep 2011. 12 pp. e-Print: arXiv:1109.1603 [physics.acc-ph]
11. H. Buchholtz, *"Electrische und Magnetische Potentiafelder"*, Springer-Verlag, 1957.
12. R.Wingerson." *'Corkscrew'-A Device for Changing the Magnetic Moment of Charged Particles in a Magnetic Field"*,May1, 1961,2pp., Published in Phys.Rev.Lett. 6 (1961) 446-448;
    http://prl.aps.org/pdf/PRL/v6/i9/p446_1

13. Handbook of Mathematical, Scientific and Engineering Formulas, Tables, Functions, Graphs, Transforms, Research and Educational Association, ISBN 0-87891-521-4.

14. Olsen, L.C. Maximon, *"Photon and Electron Polarization in High-Energy Bremsstrahlung and Pair Production with Screening"*,Phys. Rev. 114 (3) (1959) 887-904.

15. H. Bethe, W. Heitler,*"On the Stopping of Fast Particles and on the Creation of Positive Electrons"*, Proc. Roy. Soc. A 146 (1934) 83-112.

16. A.Mikhailichenko,*"Fast Bunch to Bunch Intensity Regulation in the ILC Conversion Scheme with Independent Electron/Positron Sections"*, CBN 05-18;http://www.lns.cornell.edu/public/CBN/2005/CBN05-18/CBN05-18.pdf

17. E.G. Bessonov, *"Some Aspects of the Theory and Technology of the Conversion Systems of Linear Colliders"*, 15th International Conference on High Energy Accelerators, Hamburg, 1992, p.138.

18. G.Alexander *et al*, *"Observation of Polarized Positrons from an Undulator-Based Source"*, SLAC-PUB-13145, DESY-08-025, CLNS-08-2023, COCKCROFT-08-03, DCPT-08-24, IPPP-08-12, Mar 6, 2008. 4pp. Published in Phys.Rev.Lett.100:210801, 2008.

Status of Prototyping of the ILC Positron Target

J. Gronberg*, Craig Brooksby, Tom Piggott, Ryan Abbott, Jay Javedani, Ed Cook

Lawrence Livermore National Laboratory, 7000 East Ave, L-050
Livermore, CA, 94550, USA
* E-mail: gronberg1@llnl.gov

The ILC positron system uses novel helical undulators to create a powerful photon beam from the main electron beam. This beam is passed through a titanium target to convert it into electron-positron pairs. The target is constructed as a 1 m diameter wheel spinning at 2000 RPM to smear the 1 ms ILC pulse train over 10 cm. A pulsed flux concentrating magnet is used to increase the positron capture efficiency. It is cooled to liquid nitrogen temperatures to maximize the flatness of the magnetic field over the 1 ms ILC pulse train. We report on prototyping effort on this system.

Keywords: International Linear Collider, Positron Source

1. Positron Source Overview

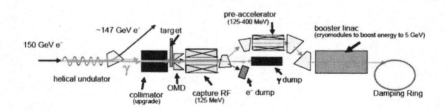

Fig. 1. A schematic layout of the ILC positron source.

These results were previously reported[1] in the Proceedings of the Linear Collider Workshop 2011. The ILC positron source[2] will be required to generate two orders of magnitude more positrons per second than any previous accelerator. As shown in Figure 1, the baseline positron system envisions passing the main ILC electron beam through several hundred meters of helical undulators in order to create a pulse of photons with energies in the

10's of MeV and over 100 kW of average beam power. The photons must then be passed through a target in order to convert a fraction of them into electron-positron pairs. This target must operate in a unique phase space compared to other target systems that have been fielded in the past. The average power that the target must dissipate is low compared to other systems but the power is concentrated into a small spot size and is deposited in a 1 ms time scale. The energy deposition in a stationary target would induce a stress in the target material which would exceed yield strength of the material and would fracture the target. The 1 ms timescale of the energy deposition makes it difficult to use motion of the target to distribute the energy deposition over a larger area. A target moving at 100 m/s will spread the energy deposition over a 10 cm stripe. In order to achieve this speed we have developed a target concept of a rotating titanium wheel that has a diameter of 1 m and rotates at 2000 RPM, as shown in Figure 2.

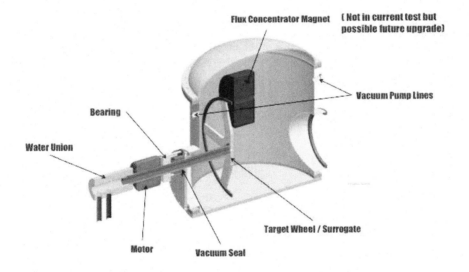

Fig. 2. A schematic of the rotating target ferro-fluidic seal test stand.

Cooling water flows through a double-walled shaft to the target where it flows out each spoke, through a section of the outer wheel, and back down a spoke. The intensity of the electron-positron beam emerging from the target makes the creation of a vacuum window to separate the target volume from the subsequent accelerator sections difficult. No design has been found which can prevent such a vacuum window from melting during operation. Therefore the target volume will share the same vacuum as the subsequent capture accelerator sections in the positron source system. In order to have the rotating target in the same vacuum as the accelerator we will need to have a rotating vacuum seal for the shaft. Rotating vacuum seals based on ferrofluids exist and are available from a number of vendors. A fluid with suspended magnetic particles exists in a gap between two counter rotating sets of permanent magnets. This forms the vacuum seal. This solution must be prototyped and studied so that the out-gassing rate from the ferro-fluid can be measured to see if a solution for the vacuum pumping can be achieved. The ability of the seal to perform continuously for the planned 9 months of operation must also be determined.

A design for a pulsed flux concentrator has also been under study. These types of devices have been used before but usually with much shorter pulse lengths. This device will need to maintain a 1 ms flat top field during the ILC bunch train. A previous device[3] created at SLAC during the 1960's for a hyperon experiment, which was designed for a 40 ms pulse, was used as the basis for the ILC design. As well as maintaining a flat top during the ILC bunch train the device will need to be able to handle the radiation environment near the target.

2. Prototyping of the Ferrofluidic Seal

Figure 3 shows a ferro-fluidic seal purchased from Rigaku corporation. The outer ring is stationary and sealed to the vacuum chamber. The inner ring has a 3 inch inner bore through which the shaft can be mounted. The ferro-fluid exists in the gaps between the two rings.

In order to test the out-gassing of the ferro-fluid into vacuum we modified an existing out-gassing test system to be able to mount the seal and rotate it at 2000 RPM. Initial commissioning of the system had problems rotating the seal at the full velocity, the drive motor would keep tripping off. Modifications were made to increase the available torque to drive the motor. The devices have a choice of the type of ferro-fluid that is used and what type of permanent magnet. We initially chose a seal with the more radiation hard permanent magnets and a more viscous ferro-fluid which should reduce out-gassing. However, a more viscous fluid also increases the torque in the system and thus the energy deposited in the ferro-fluid. While it was rated to be able to run at 2000 RPM the Rigaku sealed failed after about 15 minutes of running at 2000 RPM. It is believed that this is a heating effect and the seal was returned to Rigaku for post-mortem analysis. A second plug-compatible seal was sourced from FerroTec with a reduced

Fig. 3.   The RIGAKU ferrofluidic seal. The inner bore of three inches diameter allows the central shaft to penetrate the vacuum.

viscosity ferro-fluid for testing.

In parallel, construction of the full rotating shaft test stand is underway. The detailed design drawings for the shaft are shown in Figure 6. The shaft is composed of two concentric pipes to allow cooling water to flow to and from the shaft. A rotating water union is attached to the end of the pipe to mate with the outside water supply. We plan to use the prototype titanium wheel that was created by the University of Liverpool for eddy current testing at the Daresbury lab. Since it was not created with cooling channels the cooling water will only flow down the pipe and back. The ferro-fluidic seal is mounted on the bulkhead of the vacuum tank that we are using for this test. Farther down the shaft is a bearing block to provide support for the shaft and target. A Siemens hollow shaft motor completes the assembly and will rotate the shaft at 2000 RPM.

As of Posipol 2011 we had commissioned a vacuum tank as shown in Figure 5 and had received all of the manufactured parts for assembly of the shaft. Commissioning of a data acquisition and slow control system was

Fig. 4. A test stand to do out-gassing studies of the ferro-fluidic seal while rotating at 2000 RPM.

Fig. 5. The vacuum tank at LLNL being used for the rotating seal test.

ongoing. The long term testing will continually monitor cooling water flow rates to the ferro-fluidic seal and Siemens motor. Temperature monitors will be in place on the ferro-fluid seal, bearing block and motor as well

66

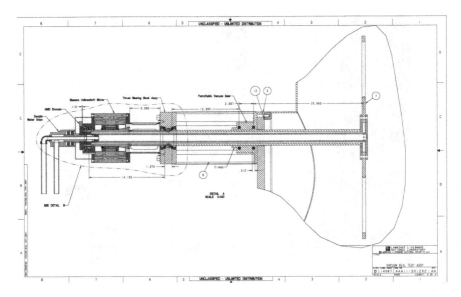

Fig. 6. Assembly drawings for the target shaft assembly.

as the cooling water. Three-axis vibration sensors will be mounted on the
ferro-fluid seal, bearing block and motor. We will use this to monitor any
degradation in the bearings over time.

3. Prototyping of the Pulsed Flux Concentrator Magnet

A pulsed flux concentrator (PFC) is basically a transformer where a set
of energizing coils induces a current in a set of concentrating plates. Figure 7 shows the six concentrating plates interleaved with the five energizing
coils. Figure 8 shows an example of one of the concentrating plates. The
insulating gap that runs from the bore to the edge is critical to successful operation. Without a gap the induced currents in the plate would have
the opposite sense as the currents in the coils and would cancel the magnetic field generated from the coils. The insulating gap from the bore to
the outer edge prevents current from circulating around the outside edge
and forces the current to travel around the bore in the same sense as the
coils. This has the effect of concentrating the field from the coils into a
smaller cross-sectional area and creating a higher field in the bore. One of
the limitations of this technique is that currents in the plate will dissipate
over time due to ohmic resistance losses in the plate reducing the field. On
a long enough timescale the currents in the concentrating disk will fall to

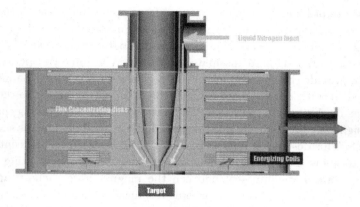

Fig. 7. Schematic of the pulsed flux concentrator. Energizing coils induce a current in the concentrating plates to create the magnetic field in the bore. The entire assembly is bathed in liquid nitrogen to reduce the electrical resistance of the plates and help maintain the current over the 1 ms ILC bunch train.

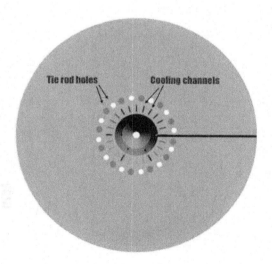

Fig. 8. One of the flux concentrating disks. An insulating layer forces the induced current to flow around the bore.

zero and the field will return what would be induced by just the energizing coils. For this application we would like to have a reasonably constant field over the 1 ms pulse length of the ILC bunch train. Therefore we plan to use OFHC copper and cool it in a liquid nitrogen bath in order to decrease it's resistance and increase the usable length of the pulse.

The site of highest energy deposition from operation of the device is in the concentrating plates near the bore of the magnet where the currents are highest. The first concentrating plate dissipates about 300 W and the second concentrating plate dissipates about 500 W at full 5 Hz operation. In order to provide the best cooling in this region a set of cooling channels are designed to run along the bore from the back end of the device to the front and then discharge into the larger liquid nitrogen bath. The cooling channels are designed as slits that are positioned radially around the bore as shown in Figure 8. This design requires an insulating seal between the concentrating plates to allow the cooling fluid to flow between the concentrating plates and to create a vacuum seal between the magnet bore vacuum and the liquid nitrogen bath. This is an extreme technical challenge as the seal must be electrically insulating, radiation hard, and able to withstand repetitive impacts from the 5 Hz pulsed operation. We have created a concept for a seal based on flexible graphite to create the radiation hard seal with a layer of Zirconia Toughed Alumina to provide the radiation hard electrical connection. Initial prototyping test will be carried out to determine if this is a viable solution.

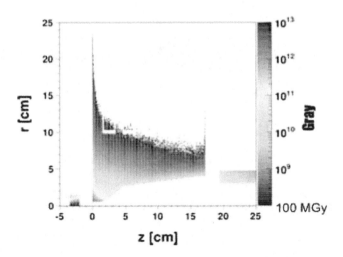

Fig. 9. A Monte Carlo simulation of radiation deposition in the material of the pulsed flux concentrator from DESY/Zeuthen over 9 months of ILC running. The 100 MGy line is the limit for using organic insulators. All insulation of the plates at higher radiation levels must use ceramics.

The creation of a liquid tight cavity for the nitrogen bath also requires that the gap in the concentrating plate be filled with a radiation hard electrical insulator that can be bonded into place to form a liquid tight seal. This bond will be a site of high stress during the pulse. The requirement that this bond survive 100 million pulses over the 9 months of operation leads to a design choice of Zirconia Toughed Alumina. This material should be able to survive the repetitive stress. This may be difficult to manufacture and we are searching for vendors who can do this type of braze.

The device sits a few centimeters from the target and is exposed to the radiation field from the device. Electrons, positrons and photons will emerge from the target and hit the device. Figure 9[4] shows the expected dose in the device for 9 months of operation as a function of position. At the front face of the device, which is directly exposed to the charged particle flux, the dose can be TeraGray. The solid copper material of the disks provides for self-shielding of the charged particles and dose falls off rapidly. However the photon flux from the target is harder to shield against. Photon conversions will provide a radiation flux at all distances falling off as distance squared and from the shielding effect of material in the path of the photon. The energizing coils and concentrating plates will need to be electrically insulated from each other. The organic insulator with the best radiation survival is Kapton which is rated out to 100 MGray. From Figure 9 we can see that it may be usable as an insulating material on the energizing coils but is unusable anywhere closer to the bore of the device.

4. Future Work

As of Posipol 2011 the out-gassing test stand for the ferro-fluidic seal was being commissioned and the full rotating shaft test stand was under construction. Assembly drawings for the pulsed flux concentrator were being started and a set of prototype tests of the seals for the liquid nitrogen containment were planned. Once component testing is complete we will proceed to create a pulser to drive a prototype magnet at full current but reduced repetition rate and create a prototype magnet with the important features that will be required for realization of a final device at ILC.

Acknowledgments

The author is grateful for the assistance of the ILC positron group in these efforts. Particularly the DESY/Zeuthen group, Sabine Riemann, Andriy Ushakov and Friedrich Staufenbiel for their calculations of radiation damage in the pulsed flux concentrator and the ANL group, Wei Gai and Wanming Liu for calculations of the capture efficiency of the pulsed flux concentrator. Thanks to Ian Bailey for the loan of the prototype titanium target wheel. This work performed under the auspices of the U.S. Department of Energy by the Lawrence Livermore National Laboratory under Contract DE-AC52-07NA27344

References

1. J. Gronberg *et al.*, "Proceedings of Linear Collider Workshop 2011", arXiv:1203.0070v1 [physics.acc-ph]
2. N. Phinney, N. Toge, N. Walker *et al.* "ILC Reference Design Report Volume 3 - Accelerator", arXiv:0712.2361v1
3. H. Brechna, D.A. Hill, and B.M. Bailey, Rev. Sci. Instrum. **36**, 1529 (1965)
4. Andriy Ushakov, DESY/Zeuthen, private communications.

HEAT LOAD AND STRESS STUDIES FOR AN DESIGN OF THE PHOTON COLLIMATOR FOR THE ILC POSITRON SOURCE

F. STAUFENBIEL*, S. RIEMANN,

DESY, Platanenallee 6, Zeuthen, D-15738, Germany

O.S. ADEYEMI, V. KOVALENKO, L. MALYSHEVA, A. USHAKOV,
G. MOORTGAT-PICK

*II.Insitut für Theor. Physik, University of Hamburg, Luruper Chaussee 149, Hamburg,
D-22761, Germany*

G. MOORTGAT-PICK

DESY, Notkestr.85, Hamburg, D-22761, Germany

The ILC baseline design for the positron source is based on radiation from a helical undulator to produce positrons in a thin target. Since the photon beam created in the helical undulator is circularly polarized, the generated positron beam is longitudinally polarized. Using a photon collimator upstream the positron target the positron polarization can be enhanced. However, the photon beam intensity yields a huge thermal load in the collimator material. In this paper the thermal load and heat dissipation in the photon collimator is discussed and design solutions are suggested.

Keywords: ILC collimator; material stress; high power beam absorption, heat loads.

1   Introduction

The ILC project proposes a high-luminosity electron-positron collider with cms-energies up to 1TeV [1]. The electron beam will have a polarization larger than 80%. The positron source design is based on a helical undulator [2] passed by the high-energy electron beam to radiate circularly polarized photons. The photon beam hits a thin Ti-alloy target and produces pairs of longitudinally polarized electrons and positrons. A photon collimator upstream the target cuts that part of the photon beam with the lower average polarization. With increasing photon beam collimation the energy deposition in the collimator material increases. Due to the time structure of the intense photon beam the peak energy deposition density (PEDD) could exceed the limits accepted by the collimator material. The material limits are given by the elastic yield strength (corresponding to the thermal activation energy of diffusion) and by the fatique stress (corresponding to the thermal recrystallization energy) over a long-term running. The limits can be estimated by temperature loads with 70% and 40% of the material melting points, respectively (e.g. see [3]). Therefore, it is necessary

---

* E-mail: friedrich.staufenbiel@desy.de

to find a proper design and material choice to prevent a collimator system breakdown. This paper presents the results of a design study for a photon collimator system for the ILC positron source. This study is based on an undulator design suggested for cms-energies up to 500GeV. In section 2 the parameters of the positron production system are given. In section 3 the heat load, the choice of the collimator material are discussed and in section 4 its dissipation; the cooling requirements are included. The last section summarizes and presents an outlook.

2   Production of polarized positrons by helical undulator radiation

The parameters of the electron drive beam are given in reference [1]. Here we consider an electron beam energy of about 250GeV, 2×10^{10} electrons per bunch and 2625 bunches per train with a train length of about 969μs and 5Hz repetition rate as suggested in the SB2009 proposal [4]. Our results correspond to the parameters presented in the Technical Progress Report 2011 [5] by scaling.

In order to produce circularly polarized photons the electrons pass a helical undulator with K=0.92, λ_0=11.5mm, located at a distance of about 500m to the positron target and photon collimator respectively. The effective undulator length is L_{und} =45m to generate the required number of positrons with a 250GeV electron drive beam. The resulting average photon beam power is P_γ=164kW.

The opening angle of the radiated photon beam is determined by the energy of the electron beam; it is proportional to $1/\gamma$. The opening angle of the higher harmonics cone is K/γ.

The degree of photon polarization depends on the emission angle and on the fractional energy of the photons (see [6]). By cutting the outer part of the radial symmetric beam, the polarization increases by contemporaneous decreasing positron yield. For the above mentioned parameter set, Table 1 shows the degree of positron polarization and the production yield for three selected collimator apertures (see also [7]).

The intensity of the undulator radiation has the maximum around the beam axis. For the given parameters, a positron polarization of 60% can be achieved for photon beam radii collimated to r=1.0mm. With larger collimation radii the positron polarization approaches about 27% for r≥3.0mm (see [8]). However, to provide the required number of positrons also for high values of polarization, a more intense photon beam is required resulting in a substantially increased heat load in the collimator material due to the increased photon beam power. For example, for 60% positron polarization the photon beam power is twice as much as for 27%.

Table 1. Polarization degree and positron yield for different collimator apertures.

collimator aperture	P_{e+}	e^+ Yield
no collimator	27%	100%
2.0 mm	35%	91%
1.4 mm	47%	73%
1.0 mm	60%	50%

3 Heat load studies for photon collimators

The aperture of the collimator determines the average polarization of the photon beam and hence, of the positrons produced in the titanium target and captured [8]. The heat load in the collimator and target is calculated using the FLUKA Monte Carlo code for particle tracking and particle interactions with matter [9]. By means of this simulation tool the optimization of the photon collimator design is done by quantifying the energy deposition in the collimator materials and selecting them corresponding to the tolerable increase of temperature and material stress, respectively [10]. The temperature rise is calculated by

$$\Delta T = \frac{\Delta Q}{m \cdot c_v} \quad (1)$$

where ΔQ is the energy deposition in the material (given in [J]), m the mass and c_v is the heat capacity (unit [J/kg/K]). The maximal temperature rise corresponds to the maximal PEDD.

3.1 *Previous photon collimator design*

So far, the design for the high power photon collimator was based on electron beam energies of 150GeV which leads to photon energies up to 10MeV for the first harmonic radiation. The collimator had a cylindrical shape and was segmented in a first part made of graphite and a second part made of tungsten [11], each with a length of z=9cm [12]. Providing the required numbers of positrons for the parameter set given above, an instantaneous temperature rise up to ΔT=1800K is expected in the tungsten part for an aperture of r=2.0mm. For an aperture of r=1.0mm the temperature rises up to ΔT=14000K. Therefore, the designs presented in [11,12] are not appropriate for the considered ILC beam parameter set.

3.2 *Moveable multistage collimators*

To achieve more flexibility in the polarization and yield manipulation for the positron beam, a moveable multistage collimator system with attenuating apertures is proposed as shown in Figure 4. The designs of the second and the third collimator take into account the collimation from the previous collimator in order to keep the lengths of the whole device as short as possible. The following calculations and simulations correspond to such system.

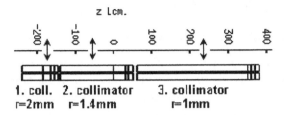

Figure 1. Moveable multistage photon collimators enable a flexible choice of the positron polarization (e.g., 27%, 35%, 47%, 60%). For this solution the required space in z direction is less than 7m.

So far, a design of moveable collimator segments each consisting of one material has been considered. The fabrication of these segments with small aperture including cooling channels has not been regarded; probably the longer collimator components will consist of partitioned segments. The proper alignment of a collimator with small aperture segments requires special care to obtain the desired reproducibility of polarization.

3.3 *Optimized collimator design*

The main fraction of energy is absorbed in the first part of the collimator. A low-Z material - pyrolytic graphite - is chosen. It's evaporation point scores up to 3650°C without a liquid phase [13]. In addition, pyrolytic graphite is very resistant against particle evaporation by energy impact. The material is strong anisotropic in the (xy) plane (basal direction) and the (z) direction. The thermal conductivity is a factor of 200 higher in the basal direction and has a very low thermal expansion coefficient [14]. However, due to the high radiation length of $X_0 \approx 19$cm, a long collimator would be needed to absorb the unwanted part of the photon beam. In order to distribute the energy deposition over a large range in the collimator material and to keep the collimator short, a proper medium-Z (or high-Z) material with smaller radiation length has to follow the graphite segment. Figures 1 to 3 show the simulated energy deposition distributions in collimators for three different apertures; the collimators are assembled with structures of pyrolytic graphite, titanium, iron and tungsten.

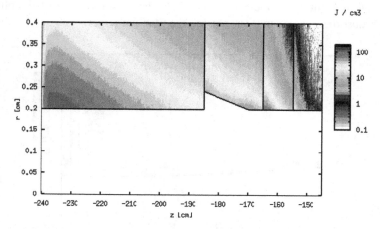

Figure 2. FLUKA simulation of the energy distribution deposited by the circularly polarized photon beam in a collimator composed of pyrolytic graphite, titanium, iron and tungsten, with an aperture 2.0mm radius and a length of z=95cm.

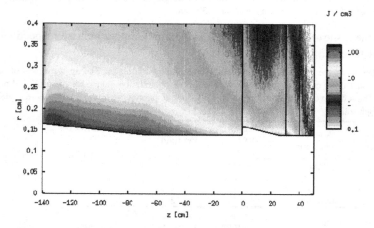

Figure 3. FLUKA simulation of the energy distribution deposited by the circularly polarized photon beam in a collimator composed of pyrolytic graphite, titanium, iron and tungsten, with an aperture 1.4mm radius and a length of z=190cm.

All unwanted photons and secondary particles are absorbed in the collimator. The lengths of the parts are optimized to lower the energy deposited in the following higher Z material to an acceptable level. The length of the last very high Z material tungsten is optimized in order to reduce the exiting shower particles below 0.1% of the absorbed photon power. For the collimator design presented in Figures 2-4 more than 99.9% of the unwanted part of the photon beam is absorbed; less than 0.1% reaches the titanium target.

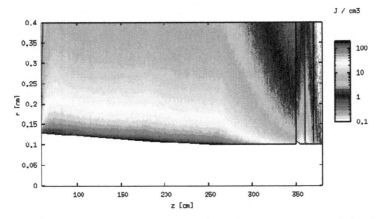

Figure. 4. FLUKA simulation of the energy distribution deposited by the circularly polarized photon beam in a collimator composed of pyrolytic graphite, titanium, iron and tungsten, with an aperture 1.0mm radius and a length of z=320cm.

Table 2 shows the PEDD and the maximum temperature rise respectively for a multistage photon collimator.

Tab.2. PEDD in the components of the multistage photon collimator. The aperture radii of 2.0mm, 1.4mm and 1.0mm correspond to 35%, 47% and 60% polarization degree for the produced positron beam. The effective undulator length is 45m, and a photon yield of 1.95ph/(e- m) yields to (4.6e15)ph/train.

Aperture r=2.0mm	E_{max} per train [J/cm^3]	ΔT_{max} per train [K]	ΔT_{max} for P_{e+}=60% L_{und}=90m per train [K]
graphite	111	70	140
titanium	56	24	48
iron	78	22	44
tungsten	31	12	24
aperture r=1.4mm	E_{max} per train [J/cm^3]	ΔT_{max} per train [K]	ΔT_{max} for P_{e+}=60% L_{und}=90m per train [K]
graphite	159	100	200
titanium	54	23	46
iron	78	22	44
tungsten	30	11	22
aperture r=1.0mm	E_{max} per train [J/cm^3]	ΔT_{max} per train [K]	ΔT_{max} for P_{e+}=60% L_{und}=90m per train [K]
graphite	176	110	220
titanium	54	23	46
iron	78	22	44
tungsten	31	12	24

The photon collimator decreases the positron yield as shown in Table 1. This has to be compensated by increasing the undulator length to $L_{und}=90m$ according to $N_\gamma \sim L_{und}$ in order to maintain the demanded luminosity at the IP. This results in a higher heat load which is taken into account in the last column of Table 2.

Due to the energy distribution in the photon beam the heat load on the material increases substantially with decreasing collimator radii. In order to distribute the load over a wide range of the collimator material, a conical shape at critical areas of high energy depositions is used. Therefore, the first sections of the graphite part from the second and third collimator and the first titanium part have a conical shape. The length of the graphite part increases strongly for smaller apertures.

The collimator design presented in Figures 1-3 implies a safety-factor to allow moderate average temperatures in the bulk materials and an efficient cooling. The dimensions of the collimators are summarized in the Tables 3-5.

3.3.1 *Dimensions of the first collimator with 2.0mm aperture*

Table.3. Dimensions of the parts for the first collimator in the whole arrangement. The total length is about 95cm.

collimator part	aperture r_1, r_2 [mm]	length z [cm]
graphite / cyl.	2.0	55
titanium / cone	2.4 / 2.0	15
titanium / cyl.	2.0	5
iron / cyl.	2.0	10
tungsten / cyl.	2.0	10

3.3.2 *Dimensions of the second collimator with 1.4mm aperture*

Tab.4. Dimensions of the parts for the second collimator in the whole arrangement. The total length is about 190cm.

collimator part	aperture r_1, r_2 [mm]	length z [cm]
graphite / cone	1.65 / 1.4	70
graphite / cyl.	1.4	70
titanium / cone	1.6 / 1.4	25
titanium / cyl.	1.4	5
iron / cyl.	1.4	10
tungsten / cyl.	1.4	10

3.3.3 *Dimensions of the third collimator with 1.0mm aperture*

Table.5. Dimensions of the parts for the third collimator in the whole arrangement. The total length is about 320cm.

collimator part	aperture r_1, r_2 [mm]	length z [cm]
graphite / 1.cone	1.3 / 1.2	60
graphite / 2.cone	1.2 / 1.1	60
graphite / 3.cone	1.1 / 1.0	80
graphite / cyl.	1.0	90
titanium / cone	1.1 / 1.0	5
titanium / cyl.	1.0	5
iron / cyl.	1.0	10
tungsten / cyl.	1.0	10

4   Radial heat dissipation for cylindrical collimators

In the equilibrium, the radial heat dissipation through a hollow cylinder with central heating is given by [15]

$$\dot{Q} = \frac{\lambda \cdot 2\pi \cdot z \cdot \Delta T}{\ln(\frac{r}{r_0})} \qquad (2)$$

where r_0 is the inner radius of the cylinder, z is the length of the cylinder, and λ the heat conductivity. Equation (2) is used to adjust the outer radius of the collimator and the cooling power required to achieve the temperature difference ΔT between inner and outer surface of the cylinder. The instantaneous heating of the inner part of the cylinder by one bunch train yields maximum temperature rise as shown in Table 2. Between the bunch trains the heat dissipates into the bulk and the temperature at the inner surface of the cylinder decreases.

Tab.6. Average maximal heat flow through a normalized area of about $1m^2$ and z=1m collimator length.

parts of the 1.collimator r_0=2.0mm	heat power through $1m^2$ girthed area [kW/m^2]
py. graphite	0.30
titanium	0.01
iron	0.02
tungsten	0.03
parts of the 2.collimator r_0=1.4mm	heat power through $1m^2$ girthed area [kW/m^2]
py. graphite	0.43
titanium	0.01
iron	0.02
tungsten	0.02
parts of the 3.collimator r_0=1.0mm	heat power through $1m^2$ girthed area [kW/m^2]
py. graphite	0.47
titanium	0.01
iron	0.02
tungsten	0.03

Assuming a homogeneous, radial directed thermal dissipation from the inner hot to the outer cooled surface, the required average cooling power corresponds to $\dot{Q}$. With heat transition coefficients of about 1kW/(m² K) as typically used for technical cooling solutions, Table 6 presents the power to be discharged per meter of collimator length through the outer surface for the radius r=16cm. The numbers in the table are related to a positron beam polarization of 60%, a bunch train length of about 1ms and a bunch train repetition rate of 5Hz.

The numbers in Table 6 should not hide the fact that the instantaneous power deposition in the photon collimator is huge; it reaches 70kW - 90kW for the parameters considered to achieve 35% - 60% positron polarization.

5 Conclusion

This multistage collimator system is a proper solution for the photon beam collimation at the ILC positron source. Due to the close correlation between the photon beam intensity, collimation and the degree of polarization the collimator system has to withstand huge heat loads without breakdown during a long operation time. The presented collimator design keeps the material loads in a comfortable regime with an additionally safety margin against failure due to fatique stress. Furthermore, the presented solution can be easily adopted to drive beam energies up to 1TeV.

Acknowledgments

We would like to thank the organizers and the host of POSIPOL 2011 for this fruitful and encouraging workshop and for hospitality.

Work supported by the German Federal Ministry of Education and research, Joint Project R&D Accelerator Spin Management, contract No. 05H10GUE.

References

1. ILC Reference Design Report (RDR), August 2007.
2. J.A. Clarke et al., Proceedings of EPAC2008 (1915), Genoa, Italy.
3. W. Schatt, Einführung in die Werkstoffwissenschaft, 4. überarb. Aufl., VEB DVfG Leipzig, 1981.
4. ILC SB2009 Proposal, 2009.
5. ILC - A Technical Progress Report, June 2011.
6. K. Flöttmann, DESY 93-161, Hamburg, 1993.
7. A. Ushakov et al., Proceedings of IPAC2011 (997), San Sebastian, Spain.
8. S. Hesselbach et al., Proceedings of PAC2009 (503), Vancouver, Canada.

9. FLUKA web site, http://www.fluka.org/fluka.php
10. WolframAlpha, computational knowledge engine, http://www.wolframalpha.com
11. A. Mikhailichenko, Proceedings of EPAC2006 (807), Edinburgh, Scotland.
12. L. Zang et al., Proceedings of PAC2009 (584), Vancouver, Canada.
13. J. Pappis and S.L. Blum, Properties of Pyrolytic Graphite, Research Division, Raytheon, Waltham, Massachusetts (592-597), 1961.
14. D. Yao and B. Kim, Applied Thermal Engineering 23, (341-352), 2003
15. W. Wagner, Wärmeübertragung, 3. überarb. Aufl., Vogel, 1991.

LITHIUM LENS FOR ILC

ALEXANDER A. MIKHAILICHENKO,

Cornell University, CLASSE, Ithaca, New York

Lithium Lens is a key element of FERMILAB proton conversion system in use for many years. We are analyzing the ILC positron source equipped with a scaled version of Lithium lens. Usage of liquid Lithium allows efficient cooling of Lithium container and entrance/exit windows. For the temperature just ~80°C higher, than the temperature of boiling water, the system for circulation of liquid Lithium is a compact and reliable. Overall efficiency of 1.5 secondary positrons per each initial electron passing the undulator is feasible with a compact Lithium lens. Axially symmetric motion of liquid Li does not perturb the field quality required for minimization of emittance of the secondary positrons/electrons polarized longitudinally.

OVERVIEW

International Linear Collider (ILC) supposed to be the next big project in High Energy Physics after LHC. Lepton collisions with theirs pure initial states, including polarization, allow much clear interpretation of results [1]. In ILC polarized electrons will collide with (polarized) positrons at 250 GeV (initially). To satisfy demands of physical community, the conversion system with undulator suggested for ILC [2]. Positron could be generated by gamma-quant in a pair with electron through the electromagnetic process involved electric field of nuclei. The gammas in ILC conversion system obtained in a helical undulator in contrast to the beamstrahlung photons of conventional system. Although the thickness of conversion target in a system with undulator can be few times less, than in conventional method, still it requires collection of positrons in a large spherical angle. A special focusing system – an optical matching device (OMD) serves for this purpose. The flux concentrator (FC) [3] which serves as OMD, accepted as the baseline for now. Advantage of FC is obvious: absence of material on the way of beams (gamma beam loses just ~15% of its intensity while passes through the target). One peculiarity of FC is that it focuses equally electrons and positrons, as its focusing properties proportional $\propto 1/e^2 \int B^2 dz$. So when the particles are focused by the flux concentrator or solenoidal lens, the total charge of beam (mixture of electrons and positrons) appeared in the first accelerator structure is about zero. So while the positrons are accelerated in the first sections of RF structure, the electrons are decelerated there. This requires adequate attention. Another peculiarity- for ILC the pulse duty is ~ 1ms, so the skin-depth phenomenon manifests itself here, forcing to make correction of feeding pulse for compensation of weakening of the focusing field during the pulse.

It is interesting to mention that in E-166 experiment, for collection of ~8 MeV positrons the DC solenoidal lens was used successfully [4]. In principle it will be not a problem to enhance its parameters so the lens will be able to collect 15 MeV particles.

One obstacle here is that the stray fields might interact with the target as it is just a Titanium rid, spinning in a close vicinity of the edge of the lens. The stray fields of the OMD in addition to the perturbation of emittance might add to the friction of the rid spinning in a stray field [5]-[6].

Lithium lens, which is basically a Lithium rod with a current co-propagating with the beam in the axial direction–is another possible candidate for OMD. However, as the beams are going through the body of Lithium, confined in a container with the input and output windows (flanges), the energy deposition in the Lithium material and in the windows is a primary point of concern here. On the other hand, the Lithium lens focuses the only particles with desired sign of charge (positrons), while the particles with other sign of charge became defocused. So the Lithium lens might serve as a preliminary energy separator; the particles with low energy became over-focused and captured in a collimator located after the lens in front of an accelerating structure.

THE LITHIUM LENS CONCEPT

The Lithium lens concept is known for a long time now [7]-[9] and a conceptual layout of Lithium lens is given in Fig. 1 here. If a steady current I runs through the rod having radius a, which axis runs along z, the azimuthal magnetic field inside the rod could be described as

$$H_{\vartheta}(r) = \frac{0.4\pi I r}{2\pi a^2} \tag{1}$$

where the magnetic field is measured in kGs, a –in cm, I –in kA. The current density comes to $j_s = I / \pi a^2$. A particle, when passed through the rod having the length L, cm, in z-direction will get the transverse kick, which is linearly dependent of transverse offset from the axis of the rod r

$$\alpha \cong \frac{\int_0^L H(r,z)\cdot dz}{(HR)} \cong \frac{0.2 I L r}{a^2 \cdot (HR)}, \tag{2}$$

The last estimation is valid for $r \ll L$ i.e. when it is possible to neglect the change of particle offset while it is running through the lens (short lens approximation).

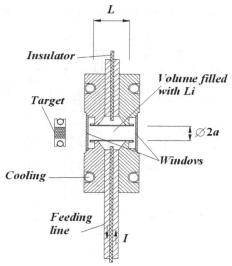

Figure 1. The Lithium Lens concept. The windows made on Be, Ti, BN or BC.

So the focal distance could be defined as the following

$$F = \frac{r}{\alpha} \cong \frac{a^2 \cdot (HR)}{0.2 I L} \tag{3}$$

For a particle with energy $E=20\ MeV$, $(HR) \approx 67\cdot\ kG\cdot cm$, $I=100kA$, $a=0.7cm$, $L=1\ cm$ the focal distance comes to a $F \cong 1.7cm$. These numbers demonstrate the possibilities of this type of focusing. Angular scattering in a material of target

$$\bar{\vartheta} \cong \frac{21MeV}{EMeV}\sqrt{\frac{L}{lX_0}} \cong \sqrt{\frac{1}{152}} \cong 0.08 rad \tag{4}$$

should be compared with the angular spread inside a secondary beam under focusing (collecting) which is ~0.5 rad. Here lX_0 is the length corresponding radiation length; for Lithium it is ~154 cm, see Table 1. Scattering in the flanges of container with Lithium is $\bar{\vartheta} \cong 0.03 rad$ for the flanges made from Boron Carbide B$_4$C with thickness 2 *mm*. Of cause, these numbers are given for estimation only. The numerical codes developed for modeling conversion of gammas into positrons take this effect of multiple scattering into account [9]. The last code KONN calculates the energy deposition and temperature gain in Lithium and flanges made from Beryllium.

Table 1. Properties of Lithium[1], Beryllium, Boron Carbide (BC), Boron Nitride (B$_4$N), and Tungsten

	Units	Li	Be	BN	B$_4$C	W
Atomic number, **Z**	-	3	4	5/7	5/6	74
Yong modulus	GPa	4.9	287	350-400	450	400
Density, **r**	[g/cm^3]	0.533	1.846	3.487	2.52	19.254
Specific resistance	Ohm-cm	1.44 x10^{-5}	1.9 x10^{-5}	>10^{14}	7.14 x10^{-3}	5.5 x10^{-6}
Length of X0, lXo	cm	152.1	34.739	27.026	19.88	0.35
Boil temperature	°C	1347	1287	Sublim. at melt	3500	5660
Melt temperature	°C	180.54	2469	2973	2350	3410
Compressibility	cm^2/kg	8.7 x10^{-6}	9.27 x10^{-7}	1.2 x10^{-6}		2.93 x10^{-7}
Grüneisen coeff.	-					2.4
Speed of sound (long)	m/sec	6000	12890	16400	14920	5460
Specific heat	J/g°K	3.6	1.82	1.47	0.95	0.134
Heat conductivity	W/cm/°C	0.848	2	7.4	0.3-0.4	1.67
Thermal expansion	1/°C	4.6x10^{-6}	11x10^{-6}	2.7x10^{-6}	5x10^{-6}	4.3x10^{-6}

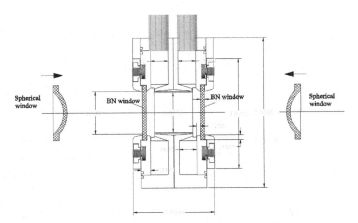

Figure 2. Cross section of the lens with liquid Lithium. Optional spherical windows serve for compensation of spherical aberrations. Dimensions in *cm*. Input/ output tubes ducting the liquid Lithium marked by a blue color.

---

[1] Total mass of Lithium in ~70*kg* human body is ~7*mg*.

For example, for K=0.92, undulator length $35m$ is enough for generation of 1.6 positrons per each initial electron in undulator; the temperature gain in Be entrance window is ~39°C at max, in the Litium ~15°C at max and in the exit window the temperature jump is ~20°C for the beam train   with 10^{13} initial electrons passed through the undulator at energy 150 GeV. Period of undulator was taken 1.15 cm.

1. COMPARISON WITH FERMILAB LENSES

Lenses with solid Lithium are in use at FERMILAB for a long time. Total amount of lenses fabricated around 24 [10]. The lenses for positrons and for antiprotons are represented in Fig.3. One can see, that the lens for positrons is much more compact one.

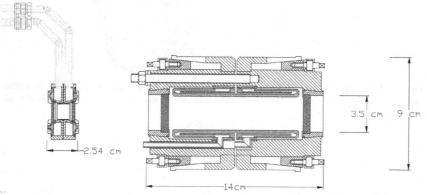

Figure 3. Lens for positrons and for the FERMILAB proton conversion target stations in comparison.

Parameters of lenses are represented in Table 1.

Table 1. Parameters of lenses for positrons, antiprotons anf fpr Neutrino-Factory

	Positrons	Antiprotons	Neutrino Factory
Diameter, cm	1.4	3.6	1.8-6
Length, cm	1	10	15
Current, kA	<75	850	500
Pulse duty, $msec$	~2	0.1	~1
Repetition rate, Hz	5	0.7	0.7
Resistance $\mu\Omega$	32	50	27
Gradient, kG/cm	<35	55	45
Surface field, kG	43	100	80-40
Pulsed Power, kW	~360	36000	6750
Average Power, kW	~3.8	3.6	4.7
Temp. gain/pulse, $°K$	45	80	80
Axial pressure, atm	19	400	256-64

SOME TECHNICAL DETAILS

More or less detailed description of lens one can find in Refs [10]-[13]. The latest design of the lens with liquid Lithium is represented in Fig. 4 below. The latest design of the combined heat exchange/pumping device with the gear pump for the liquid Lithium is represented in Fig.5.

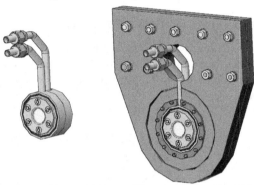

Figure 4. At the left: The lens with liquid Lithium. Diameter of lens ~2*in*, thickness~1*in*. The tubing serves for the Lithium in/out. The cross section is represented in Fig.2. At the right: The lens installed into the current duct.

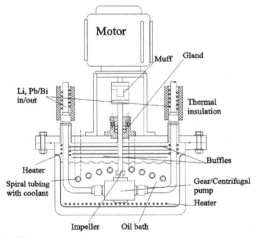

Figure 5. Pumping and cooling system.

High flash point of oil[2] ~300°C allows normal operation of this system. In Fig. 6 two possible options for the current duct are represented in comparison. The whole conversion unit is represented in Fig. 7.

---

[2] 561® Transformer Fluid, for example. Fire poit for this liquid >340°C.

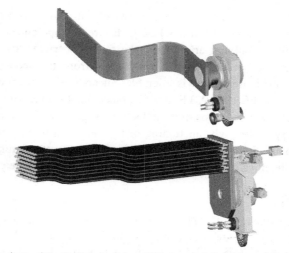

Figure 6. Two variants of the feeding duct: the stripline-at the left, with coaxial cables- at the right. Just a tip of the Lithium lens is visible in these pictures (see Fig.3).

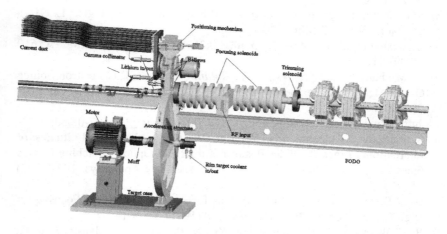

Figure 7. Positron conversion system assembled. This is a variant with a spinning target rim.

SUMMARY

Collection of positrons with Lithium lens rises evident question about possibility its components (windows mostly) against severe exposure by secondary beams of positrons and electrons. Calculations show however that the temperature gain is tolerable. The heat removal with liquid Lithium flow is

88

adequate. Some magneto-effect of interaction of Lithium flow with magnetic field [12] helps in mixture of Lithium having different temperatures. The general temperature gain in Lithium arises from the feeding current flow. Indeed the maximum input into heating in windows arises from the secondary beams. Anyway the temperature regime is tolerable. Successful experience of Lithium lens exploitations at FERMILAB brings assurance that the lens with Lithium is a feasible device. In a case of positron production usage of *liquid* Lithium is much more beneficiary. Although functioning of the lens with liquid Lithium is not tested experimentally yet, all parameters of this system remain within guaranteed by the physical properties of all materials and components involved.

REFERENCES

1. G. Moortgat-Pick, et al., "The Role of Polarized Positrons and Electrons in Revealing Fundamental Interactions at the Linear Collider", Published in Phys. Rept. 460 (2008) 131-243, e-Print: hep-ph/0507011
2. J.A.Clarke et al., "The Design of the Positron Source for the International Linear Collider EPAC08-WEOBG03, Jun 25, 2008. 3pp., Conf. Proc C08-06-23.3 (2008).
3. J.Gronberg, T.Piggott, J.Javedani, R.Abbott, C.Brown, L.Hagler, LLNL, a Talk at LCWS10, Beijing, 2010, 27pp.
4. G.Alexander et al., Phys.Rev.Lett.100:210801 (2008); NI&M A 610 (2009) 451–487.
5. A. Mikhailichenko, *Issues for the Rotating Target*, ILC Positron Source Group Meeting September 27-28, 2006, RAL, UK; published as CBN-07-02. 2007.
   http://www.lns.cornell.edu/public/CBN/2007/CBN07-2/cbn07-02.pdf
6. S. Antipov, L. Spentzouris, W. Liu, W. Gai , "Numerical Studies of International Linear Collider Positron Target and Optical Matching Device Field Effects on Beam" , Jul 2007. Published in J.Appl.Phys. 102 (2007) 014910
7. B. F. Bayanov et al., "A Lithium Lens for Axially Symmetric Focusing of High Energy Particles Beams", NIM 190 (1981), pp. 9-14.
8. B. F. Bayanov, A. D. Chernyakin, Yu. N. Petrov, G. I. Silvestrov, V. G. Volokhov, T. A. Vsevolozhskaya, J. MacLachlan (Novosibirsk, IYF & Fermilab) ," The Proton BeamLithium Lens for the Fermilab anti-Proton Source", FERMILAB-TM-1000, Aug 7, 1980. 31pp.
   http://lss.fnal.gov/archive/test-tm/1000/fermilab-tm-1000.pdf
9. G.Biallas, et al.,"Power Tests of the Fermilab Lithium Lens for Antiproton Collection", Proc. of XII Int. Conf. On High Energy Accelerators, Fermilab, August 11-16 1983, Proc., pp.591-593.

10. A, Mikhailichenko, "Usage of Liquid Metals in Positron Production System of ILC", Presented at FRIB, Michigan State University, East Lansing, May 23, 2011; CBN 11-2.
http://www.lepp.cornell.edu/public/CBN/2011/CBN11-2/CBN11-2.pdf

11. A.Mikhailichenko, "Lithium Lens for Positrons and Antiprotons in Comparison", CBN 09-8, October 30, 2009;
http://www.lepp.cornell.edu/public/CBN/2009/CBN09-8/CBN%2009-08.pdf

12. A. Mikhailichenko, "LthiumLens (I)", CBN-09-4. Aug 2009. 17 pp.;
http://www.lepp.cornell.edu/public/CBN/2009/CBN09-4/CBN%2009-04.pdf

13. A. Mikhailichenko, "Lithium Lens (II) Lithium Flow Magneto-Hydrodynamics", CBN 10-3, 2010;
http://www.lns.cornell.edu/public/CBN/2010/CBN10-3/CBN_10-3.pdf

14. A.A. Mikhailichenko, "Lithium Lens for Positron Production System. EPAC08-WEPP157. Jun 25, 2008. 3 pp. Published in Conf. Proc. C0806233 (2008) WEPP157.

SPIN TRACKING AT THE ILC POSITRON SOURCE[*]

V. KOVALENKO[†1], O.S. ADEYEMI[1], A. HARTIN[3], G. A. MOORTGAT-PICK[1,3],
L. MALYSHEVA[1], S. RIEMANN[2], F. STAUFENBIEL[2], A. USHAKOV[1].

[1]II. Institute for Theoretical Physics, University of Hamburg,
Hamburg, 22607, Germany

[2]II. Deutsches Electronen-Synchrotron, DESY
Zeuthen, 15738, Germany

[3]II. Deutsches Electronen-Synchrotron, DESY
Hamburg, 22607, Germany

In order to achieve the physics goals of future Linear Colliders, it is important that electron and positron beams are polarized. The baseline design at the International Linear Collider (ILC) foresees an e+ source based on helical undulator. Such a source provides high luminosity and polarizations. The positron source planned for ILC is based on a helical undulator system and can deliver a positron polarization of 60%. To ensure that no significant polarization is lost during the transport of the e- and e+ beams from the source to the interaction region, precise spin tracking has to be included in all transport elements which can contribute to a loss of polarization, i.e. the initial accelerating structures, the damping rings, the spin rotators, the main linac and the beam delivery system. In particular, the dynamics of the polarized positron beam is required to be investigated. In the talk recent results of positron spin tracking simulation at the source are presented. The positron yield and polarization are also discussed depending on the geometry of source elements.

1. Introduction

The undulator scheme of polarized positron production was proposed by Michailichenko and Balakin in 1979 [1] and has been chosen as a baseline for the International Linear Collider (ILC) (see Figure 1). The scheme is based on a two stage process, where at the first stage the circularly polarized photons are

[*] This work is supported by the German Federal Ministry of Education and Research, Joint Research Project R&D Accelerator "Spin Management", contract number 05H10GUE.

[†] e-mail: valentyn.kovalenko@desy.de

generated in a helical magnetic field and then, at the second stage, these photons are converted into longitudinally polarized positrons and electrons in a thin target. The circular polarization of the photons is transferred to longitudinal polarization of the electrons and positrons. The main parts of the positron source are: the helical undulator, the photon collimator, the target, the optical matching device installed after the target in order to capture the longitudinally polarized positrons, the RF section embedded in a solenoid to capture and pre-accelerate the beam up to 125 MeV, then the pre-acceleration to 400 MeV, and after that the booster linac accelerates the beam to 5 GeV. In order to preserve the polarization of the beam in the damping ring (DR) the spin orientation of the positrons has to be rotated from the longitudinal into the vertical direction before the damping ring via a spin rotator.

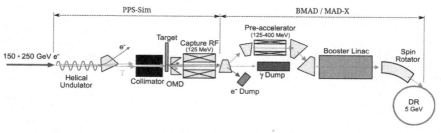

Figure 1. Schematical layout of polarized positron source based on undulator scheme.

The efficiency of the beam generation as well as the beam optics downstream the target plays the crucial role for the design of the positron source components. In this study we investigate a quarter-wave transformer (QWT) as an optical matching device in order to find the optimal geometry parameters to fulfill the beam yield requirements also for high degrees of positron beam polarization.

2. Polarized Positron Source Simulation (PPS-Sim) Code

There are several Monte-Carlo tools, for example FLUKA and EGS which are used to perform positron production simulations. But none of them allows to calculate particle beam dynamics in the accelerating structure, because the electrical field is not implemented. At the same time there are a numerous well developed codes for simulations of beam dynamics in the accelerating structures of linear and circular accelerators (MAD-X, PARMELA, BMAD, Elegant, etc.). However all these codes do not include the positron production process and require an input file from FLUKA or EGS and some of them do not take into account spin of particles.

Therefore, we used the program code PPS-Sim [2, 3] which is based on Geant4 including positron production, energy deposition and also the transport of charged particles in magnetic and electric fields, and the spin transport. PPS-Sim is an ideal tool to combine beam generation, beam focusing and particle acceleration taking into account the spin of the particles.

3. Positron Source Parameters and Simulation Results

The positron source parameters for the ILC used for the simulations are presented in Table 1.

Table 1. ILC Positron source parameters.

Electron beam energy	< 250 GeV
Number of positrons	$3 \cdot 10^{10}$ e$^+$/bunch
Number of bunches	2625 or 1312 bunches/train
Repetition rate	5 Hz
Undulator K-value	0.92
Undulator period	11.5 mm
Undulator length	231 m
Undulator-Target Distance	~ 500 m
Target material	Ti6Al4V
Target thickness	0.4 X_0
Target rotation speed	100 m/s
OMD	QWT
DR acceptance: energy spread	1 %
DR acceptance: emittance, $\varepsilon_{nx}+\varepsilon_{ny}$	0.09 rad m
DR acceptance: long. bunch size	34.6 mm

The circularly polarized photons hit Ti alloy target with a thickness of 0.4 radiation length and produce longitudinally polarized positrons. Then the generated positron beam is collected and accelerated to 125 MeV. The optical matching device is a QWT and consists of three solenoids (see Figures 2). The first two solenoids (bulking and focusing) have a higher magnetic field than the third one that is called background solenoid. The OMD is followed by a 1.3 GHz accelerating cavity embedded into solenoid with constant B-field. The E-field of the RF cavity is modeled as harmonic function. PPS-Sim does not include the whole beam line up to the DR at 5 GeV. In order to estimate the number of positrons out off the DR acceptance, PPS-Sim applies cuts on the longitudinal bunch size and on the sum of x- and y-emittances. The model of the positron source with quarter-wave capturing is shown in Figure 3.

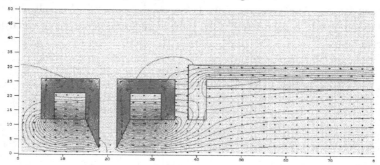

Figure 2. Magnetic field distribution in the quarter-wave transformer.

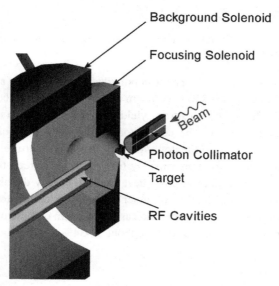

Figure 3. Model of quarter-wave transformer capturing for positron source. Only the focusing magnet and background solenoid of the QWT are shown.

3.1. Results and discussion

In Figure 4 it is shown how the polarization and yield of positrons depend on the magnetic field of QWT for different drive beam energies. In the case of a 250 GeV drive beam the polarization lies in a range of 24-26%. Different magnetic fields of the focusing solenoids of QWT (1÷2T) do not significantly affect the polarization. This value of polarization degree is corresponding to the current baseline design not sufficient to achieve the full physics potential with polarized beams. The positron yield is required to be 1.5 e^+/e^-. It should be noted that the yield is calculated as the ratio of captured positrons to electrons in the drive beam.

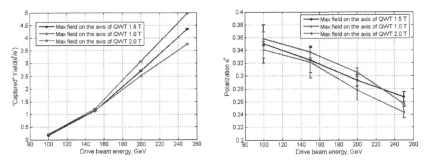

Figure 4. Polarization and yield of positrons versus different magnitudes of the QWT magnetic field for different drive beam energies.

One method to increase the positron polarization degree is to apply a photon collimator. Let us consider 250 GeV beam and how the photon collimator with different radii affects the positron yield and polarization (see Figure 5). Obviously smaller radius of the collimator results in higher polarization and lower yield. The photon collimator with 1 mm radius aperture increases the positron polarization up to approximately 60% which is agreed to be the goal. However, one has to be deal whether heat loading, energy deposition or other factors might lead to destruction of inner part of the collimator. Hence, the results presented below will consider the case with 2 mm collimator radius. For a 250 GeV electron beam a polarization of about 31-34% can then be achieved.

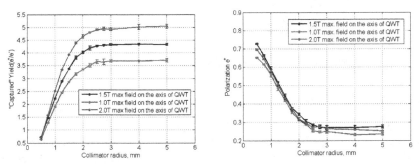

Figure 5. Polarization and yield of positrons versus different radii of the photon collimator for a drive beam energy of 250 GeV.

Figure 6 shows the polarization and yield depending on the distance between the QWT and the target. We study the initial conditions:

- Drive beam energy 250 GeV
- K=0.92, λ=11.5 mm
- No collimation
- Distance between undulator center and QWT ~500 m
- Undulator length is 231 m
- Length of QWT is 130 mm
- Maximal magnetic field on the axis of QWT is 1 T

It should be noted that the phase of the RF field is optimized to get a higher value of positron yield. If we place the QWT at 10 mm from the target we will get a maximum yield and a minimum polarization of about 23%. Increasing the distance we lose particles and the yield goes down but at the same time polarization grows up. For example, if we place the QWT at 150 mm from the target it is possible to get 28% positron polarization still providing the yield of 3 positrons per electrons.

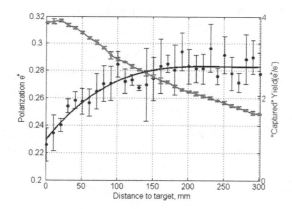

Figure 6. Polarization and yield of positrons versus distance between QWT and target without collimation for 250 GeV drive beam energy.

Using the same conditions mentioned above and applying a photon collimator with a radius of 2 mm in addition, a polarization enhancement of 6-7% can be observed. At the same time the yield still fulfils the requirement of 1.5 positrons per electrons (see Figure 7).

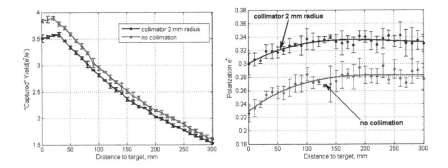

Figure 7. Polarization and yield of positrons versus distance between QWT and target with 2 mm radius of photon collimation for 250 GeV drive beam energy.

We also changed the length of the QWT in our simulations. In Figure 8 the corresponding dependencies are presented. The photon collimator was not applied in this case. It can be observed that the polarization only increased up to 24.5% caused by lengthening of QWT. The optimum yield is achieved for QWT lengths of 110 – 120 mm. The polarization in this case is 23%.

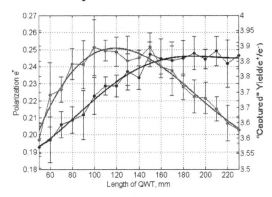

Figure 8. Polarization and yield of positrons versus length of QWT without collimator for 250 GeV drive beam energy.

4. Summary

For a 250 GeV electron beam and without photon beam collimation, the positron polarization lies in a range of 24-26%. This value of polarization might not be sufficient to achieve the full physics potential of the ILC with polarized beams. Increasing the distance between the target and the QWT gives an enhancement of polarization by 6-7%. A photon collimator with 2 mm aperture radius increases the polarization up to 35% where the positron yield still fulfils the requirement of 1.5 e+/e-. A longer QWT also slightly increases the value of polarization in addition.

References

1. V. E. Balakin, A. A. Michailichenko, *The conversion system for obtaining high polarized electrons and positron*, **INP 79-85**, Novosibirsk (1979).
2. A. Ushakov, S. Riemann and A. Schaelicke, *Positron source simulations using Geant4*, **THPEC023**, (Proceedings of IPAC2010, Kyoto, Japan), (2010).
3. PPS-Sim web site, http://pps-sim.desy.de.

THE TRULY CONVENTIONAL POSITRON SOURCE FOR ILC

TSUNEHIKO OMORI[1], JUNJI URAKAWA

KEK: High Energy Acceleratoor Research Organization, 1-1 Oho, Tsukuba-shi, Ibaraki 3050801, Japan

TOHRU TAKAHASHI, SHIN-ICHI KAWADA

Graduate School of Advanced Sciences of Matter, Hiroshima University, 1-3-1 Kagamiyama, Higashi-Hiroshima, 739-8530, Japan

SABINE RIEMANN

Deutsches Elektronen-Synchrotron, DESY, Platanenallee 6, D-15738 Zeuthen, Germany

WEI GAI, WANMING LIU

Argonne National Laboratory, 9700 S. Cass Avenue Argonne, IL 60439, USA

JIE GAO,   GUOXI PEI

Institute of High Energy Physics, 19B YuquanLu, Shijingshan District, Beijing, 100049, China

NATSUKI OKUDA

Department of Physics, Graduate School of Science, The University of Tokyo, 7-3-1 Hongo, Bunkyo-ku, Tokyo 113-0033, Japan

ANDRIY USHAKOV

University of Hamburg, Luruper Chaussee 149, D-22607 Hamburg, Germany

---

[1] Corresponding author. Tel.:+81-29-8645370; fax: +81-29-8642580
E-mail address: tsunehiko.omori@kek.jp

We propose the conventional positron source driven by a several-GeV electron beam for ILC. Thermal load of the positron production target was a risk of the conventional positron source. To cure it, we employ a 300 Hz electron linac to create positrons with stretched pulse length. In ILC, the bunch timing structures and pulse timing structures can be different in the positron source, in the DR, and in the main linac. We have some flexibility to choose timing structures in positron source and we use it for time stretching. ILC requires about 2600 bunches in a train in the main linac which pulse length is 1 ms. In the conventional source, about 130 positron bunches are created by each pulse of the 300 Hz linac. Then 2600 bunches are created in 63 ms. We optimized parameters such as drive beam energy, beam size on the target, and target thickness to maximize the capture efficiency and to mitigate the target thermal load. A slow rotating tungsten disk is employed as positron production target.

1. Introduction

The superconducting RF acceleration technology in ILC main linacs allows to accelerate high current beams. The beam pulse has 1ms duration and contains about 2600 bunches of positrons (electrons). Each bunch contains 2×10^{10} positrons (electrons). With such high current ILC will realize the high luminosity, 2×10^{34} cm^{-2}s^{-1} at E_{CM} = 500 GeV.

On the other hand, the high current gives us very challenging issue of designing the positron source. One of biggest risk areas in ILC is the thermal load on the positron production target.

The baseline choice of the ILC positron source is the helical undulator scheme. The undulator scheme gives interconnection to nearly all sub-systems of the ILC, because it uses electron beam in the electron main linac to produce positrons. A large-scale interconnected system is a challenge and strict constraints are given to the positron source and the target heat load by the time structure of the beams. In the undulator scheme, we are constrained to create 2600 bunches in 1ms. However, the helical undulator scheme provides a polarized positron beam. This advantage will be essential when the LHC will find physics scenarios which can be studied with higher sensitivity if both beams of a electrons-positrons collider are polarized [3].

On the other hand, the LHC measurements may give strong hints to physics scenarios where positron polarization would add only marginal information. Then a conventional unpolarized positron source could be sufficient.

The experiences are big advantage of employing a conventional positron source. Up to now, only the conventional positron generation scheme has been experienced in real accelerators [4]. The experiences makes us to control risks in a limited area, that is the target area. However, if polarized positrons are required in a later stage of the experiments, we need to replace the positron source.

The proposed positron source contains risks only in the target area. Therefore, we concentrate to cure these risks in two ways: (1) pulse stretching by

300 Hz generation; the proposed scheme creates 2600 bunches in about 60 ms, and (2) optimized drive beam and target parameters. Considering the choice of the target material, there are several possible options of target material, for examples, liquid metal, crystal, crystal-amorphous hybrid, and ordinary solid target. Also we have a choice of number of targets, single or multiple target. In the view point of risk control of the ILC project, we choose a single target made of solid metal because it is the most experienced scheme.

2. The 300 Hz positron generation for the time stretching

We employ stretching of the pulse length of a bunch-train in positron generation to ease target thermal and shockwave issues. The schematic view of the 300 Hz scheme is shown in Fig.1. The point is that, in ILC, the bunch timing structures and pulse timing structures can be different in the positron source, in the DR, and in the main linac. So we have some flexibility to choose timing structures in positron source and we use it for time stretching. The repetition rate of ILC main linac is 5 Hz, therefore there in an interval of 200 ms between two pulses. This gives us enough time for pulse stretching. We employ a normal conducting 300 Hz electron linac to create positrons. The pulse to pulse separation of the linac is 3.3 ms. Each pulse of the 300 Hz linac creates about 130 bunches, so 20 linac pulses create 2600 bunches of positron in about 60 m seconds.

As illustrated in Fig. 2, for the timing structure of the conventional source, in addition to the pulse stretching, we consider the matching to the timing structure of the beam in the damping ring. The bunch-to-bunch separation of 6.15 ns in the 300 Hz linac is chosen, because the bunch-to-bunch separation in the damping

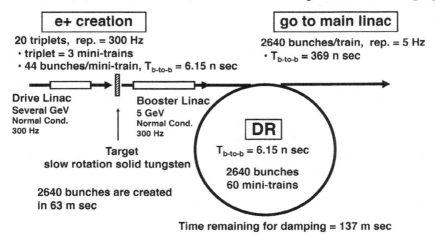

Figure 1. Schematic view of the 300 Hz positron generation scheme. The repetition rate of the drive linac and the booster linac is 300 Hz, whereas the repetition rate of the damping ring and main linac is 5 Hz.

ring is 6.15 ns. In addition, in the damping ring the positron beam has a mini-train structure where about 40 bunches form a mini-train. There are the gaps of about 100 ns between adjacent mini-trains to prevent the instability caused by electron clouds. To match the beam structure of the damping ring, in the 300 Hz linac, one RF pulse accelerates three mini-trains with inter-mini-train gaps. This package of three mini-trains is named triplet in this article. Each mini-train contains 44 bunches. With 6.15 ns bunch-to-bunch separation and 44 bunches per mini-train, the length of a mini-train is 264 ns. Since a triplet contains three mini-trains, it consists of 132 bunches and we need 20 triplets to form 2640 bunches. There are gaps of about 100 ns between the mini-trains in a triplet. Since the triplets are produced by the electron beam they have a repetition rate of 300 Hz. Therefore, we have 3.3 ms between two triplets.

As a whole the beam parameter assumed in this article is the ILC nominal beam parameter, but with slight modifications. In the original ILC nominal parameter set, one RF pulse in the main linac has 2625 bunches. Here, we assume 2640 bunches per pulse.

A positron production target made of tungsten alloy was assumed. Multiple targets, employing 3 or 4 targets, are not necessary. We assume that the target is

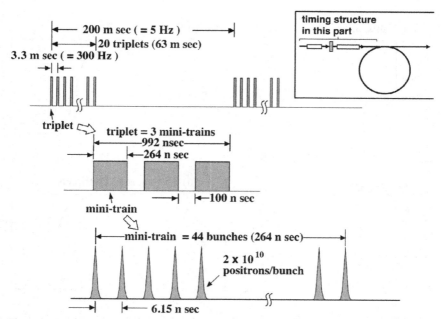

Figure 2. The bunch/train timing structure in the positron source and in the booster linac. After the booster linac, the kicker with pulse length of about 1 μs and repetition rate 300 Hz is employed to send the triplet to the damping ring.

rotating. Since the target is rotating, different triplets hit different parts of the target. The target does not need to survive the impact of 2640 bunches at the same spot. The instantaneous temperature rise caused by successive triplets impinging on the target is simulated. The results are shown in Figure 3. The tangential speeds assumed are (a) 5 m/s, (b) 2 m/s, (c) 1 m/s, and (d) 0.5 m/s. A bunch charge of 3.2 nC and a drive beam energy of 6 GeV are assumed. The spot size of the drive beam on the target is 4 mm (rms) and the target thickness is 14mm. The choices of the drive beam parameters and of target thickness are discussed later.

As shown in Figure 3d, when we choose rotation speed of 0.5 m/s, the temperature rise of the target is about 1300 K. This is significantly lower than the melting point of tungsten, 3697 K. Therefore a tangential speed of 0.5 m/s is sufficient to tolerate the heat load.

The slow rotation speed is an advantage of the 300 Hz conventional positron source. The rotation speeds of the target of the conventional positron source assumed in this article are much smaller than that of ILC baseline positron source. In the baseline positron source based on the undulator scheme, the rotation speed of 100 m/s of the target is assumed. In the simulation, of the conventional source, we did not consider cooling of the target. The analysis of the equilibrium

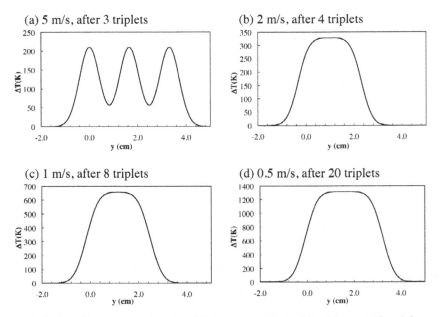

Figure 3. Instantaneous temperature rise of the target caused by the hits of the successive triplets. The assumed tangential speeds are (a) 5 m/s, (b) 2 m/s, (c) 1 m/s, and (d) 0.5 m/s. In the simulation, a bunch charge of 3.2 nC and a drive beam energy of 6 GeV are assumed.

temperature with a realistic target design which include cooling is a future plan.

At the downstream of the target, there is a positron capture system consists of an Adiabatic Matching Device (AMD) and a RF-section. The RF-section is a L-band accelerating structure. We assume a flux concentrator as an AMD. In the 300 Hz scheme, the pulse length of the triplet is about 1 μs. Therefore pulse length of the flux concentrator is also about 1 μs. The pulse length, 1 μs, is similar to that of existing flux concentrators. So the technology already exists. Further, the pulse length is short enough to use a high acceleration gradient of the RF-section. We will discuss the details of the capture system later.

After exiting the capture section, the positron energy is boosted to 5 GeV in a 300 Hz normal conducting linac. Then kicker with pulse length of about 1 μs and repetition rate 300 Hz is employed to send the positrons to the DR. One kicker pulse sends a triplet to DR. The kicker with 1 μs pulse length can be build with existing technology

After the damping, each bunch is extracted from the DR by the fast kicker, sent to the bunch compressor, and sent to the main linac. The fast kicker has a capability of bunch-by-bunch extraction. This part remains the same as in the ILC

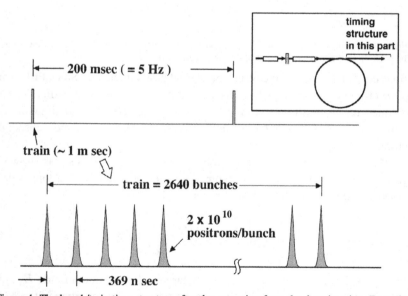

Figure 4: The bunch/train time structure after the extraction from the damping ring. From the damping ring each bunch is extracted by the fast kicker. The fast kicker has a capability of bunch-by-bunch extraction.

baseline design with an undulator positron source. The bunch-to-bunch separation is 369 ns after the extraction (Figure 4).

3. Choice of drive beam and target parameters

In order to reduce the peak energy deposit density (PEDD) and total energy deposit in the target, in addition to employ the 300-Hz-generation, it is important to optimize the drive beam and target parameters. We made calculations of positron yield, PEDD, total energy deposit and instantaneous temperature rise for

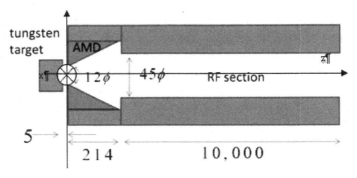

Figure 5. Layout of the target and the positron capture system implemented in the simulation. The positron capture system consists of the AMD and the RF section.

various combinations of drive beam energy and target thickness. Since PEDD and capture efficiency depend on the transverse size of the drive beam, estimations are performed for several drive beam sizes. Geant4 was used for the simulation of positron generation in the target and the particle tracking in the subsequent capture section.

The target and the capture section assumed in the analysis was shown in Figure 5. The solid tungsten was assumed as a target material. The longitudinal magnetic field of the AMD is described as

$$Bz(z - z_0) = \frac{B_0}{1 + \mu(z - z_0)} + Bsol \qquad (1)$$

,where z = 0 is the back end of the tungsten target. The maximum field (B_0) of 7 T and the taper parameter (μ) of 60.1 m^{-1} were assumed. The aperture of the AMD at the entrance was 12 mm in diameter. Between the target and the AMD, we need a gap to accommodate the rotation target. The gap, z_0, was assumed to be 5 mm. The radial field is calculated according to the prescription described in Ref [5] as

$$Br(r, z) = -\frac{1}{2} r \frac{\partial B(z)}{\partial z} + \frac{1}{16} r^3 \frac{\partial^3 B(z)}{\partial z^3}, \qquad (2)$$

where r is the transverse distance to the center of the AMD and of the RF section. After the AMD, the RF acceleration of 1.3 GHz traveling wave was applied from z = 219 to 10219 mm. The gradient was assumed to be 25 MV/m. The aperture of the RF section was assumed to be 45 mm in diameter. In addition to the AMD field, a constant magnetic field $Bsol = 0.5$ T was applied in the AMD and the RF section, i.e., from z = 5 to 10129 mm. A part of the electromagnetic shower produced in the target hits the AMD and the RF section. We calculated the energy deposit in the AMD and the RF section. In the calculation the outside of the applied fields in lateral direction, indicated by solid areas in Figure 5, was assumed to be iron while the inner area was air of 10^{-7} Pa to mimic the inside of the accelerator structure.

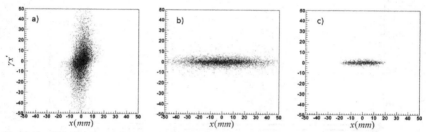

Figure 6.   The transverse phase space distribution of positrons at the entrance of the AMD (a), the exit of the AMD (b) and the exit of the RF section (c). Here we assumed the drive beam energy of 6 GeV and the beam spot size of 4 mm (rms) on the target.

We performed the simulation with various target thickness and various energy of the drive electron beam for beam sizes on the target $\sigma = 2.5$ mm and 4.0 mm. The simulation includes generation of electromagnetic shower at the target, and the tracking in the AMD and in the RF section. Figures 6 shows typical distributions of positrons in the transverse phase space at the entrance ((a) z = 5 mm), at the exit ((b) z = 214 mm) of the AMD, and at the exit of the RF section ((c) z = 10,219 mm).   The normalized emittance at the exit of the RF section was calculated as

$$\gamma \varepsilon_x = \sqrt{\langle x^2 \rangle \langle (\gamma x')^2 \rangle - \langle x(\gamma x') \rangle}. \qquad (3)$$

The twiss parameters for the distribution were obtained:

$$\alpha_x = -\langle x(\gamma x')\rangle / \varepsilon_x$$
$$\beta_x = -\langle x^2\rangle / \varepsilon_x \qquad (4)$$
$$\gamma_x = (1 + \alpha_x^2) / \beta_x .$$

Then, the parameter $A_x \equiv \gamma_x x + 2\alpha_x x(\gamma x') + \beta_x (\gamma x')^2$ was calculated for each particle, as well A_y for the y direction. The transverse acceptance of the damping ring was constraint to $A_x + A_y < 0.09mm$. The phase of the RF section was scanned to maximize the number of positrons.

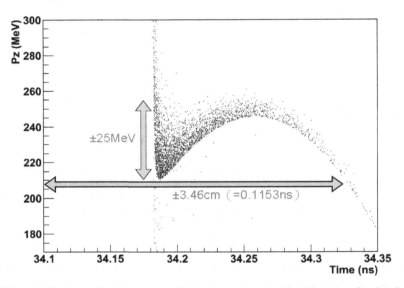

Figure 7. The longitudinal phase space distribution at the exit of the RF section after the phase optimization. The arrows show the acceptance of the damping ring. Here, the drive beam energy of 6 GeV and the beam spot size of 4 mm (rms) on the target were assumed.

A typical distribution in the longitudinal phase space at the exit of the RF section is shown in Figure 7. The damping ring acceptance limits the longitudinal phase to ±25MeV and ±3.46cm of the central values as indicated in the Figure 7. The accepted number of positrons within the damping ring acceptance for a driving electron beam size of $\sigma = 4.0$ mm is shown in the contour plot Figure 8. Also shown in the figure are the estimated peak instantaneous temperature rise and the line of the PEDD of 35 J/g which is the maximum tolerated value estimated by the previous study [5-7]. To estimate the PEDD and the peak instantaneous temperature rise, we assumed the 300 Hz scheme, 132 bunches in a triplet, and a bunch charge of 3.2 nC. We also assumed that the effect caused by all the 132 bunches is accumulated and contributes to both PEDD and peak instantaneous temperature rise.

Target rotation speed is important parameter to maximize feasibility of the positron source. We made consideration of the ration speed.

Duration of the successive bunch hits which cumulatively contributes to the target damage is not certain, therefore we estimated the damage from the largest possible deposit. To be more specific we assumed that damage due to spatial and temporal concentration of the energy deposit by 132 bunches in a triplet in 996 ns contribute equally.

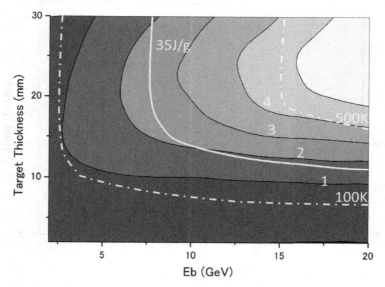

Figure 8.   Colored contour shows the number of accepted positrons per incident electron for a driving electron beam size of 4.0 mm. The lines for a PEDD of 35 J/g (solid line) and the peak instantaneous temperature rise (dot dash line) are also shown. To estimate the PEDD and the peak instantaneous temperature rise, we assumed that the contributions of the 132 bunches in a triplet were accumulated. The bunch charge of 3.2 nC was assumed. If we assume that the target will be broken when the PEDD exceeds 35 J/g, the upper-right area above the 35 J/g line is excluded.

First we discuss the temperature rise. As discussed in the previous section, the time duration of a triplet is shorter than the time of thermal diffusion. But the time between triplets (3.3 ms) is sufficiently long to achieve that each triplet hits a different position of the target rotating with tangential speed of about 5 m/s (see Figure 3a). Even if we chose the tangential speed lower than 5 m/s, the temperature rise can be acceptable. Figure 3d shows that a tangential speed of 0.5 m/s is acceptable concerning the temperature rise.

Next we discuss the effect of the thermal shockwave. With 0.5 m/s, the spatial separation of two successive triplets on the target is 1.7 mm which is smaller than the beam spot size if we employ a beam size larger than 1.7 mm in rms. Thus successive triplets have spatial overlap on the target. However, since thermal shock waves in the target develop within 1 μs or less, shock waves from successive triplets with 3.3 ms time interval should not be cumulative. Therefore the spatial overlap doesn't matter. If we use the PEDD as a measure of the effect of the shock wave, a tangential speed of 0.5 m/s is acceptable also in this respect.

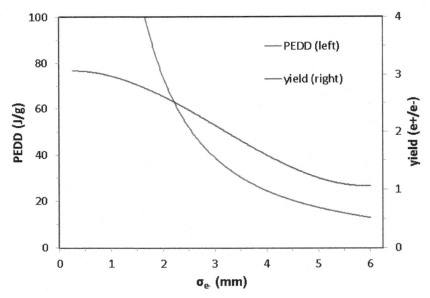

Figure 9. The PEDD (blue, left scale) and positron yield (red, right scale) as a function of drive electron beam size. Here we assume that the drive beam energy was 6 GeV, the bunch charge was 3.2 nC, and the target thickness was 14 mm. We assumed the 300 Hz scheme and 132 bunches in a triplet to make a cumulative contribution on the PEDD.

The actual choice of the tangential speed depends on further optimization and on the engineering design of the target system. We need further studies for the final design of the positron source. Figure 9 shows the positron yield and the PEDD as function of the size of the driving beam. In Table 1, the parameters of the proposed positron source design are summarized with assuming a 6 GeV drive beam of size 4 mm, which hits a tungsten target of 14 mm thickness. The average energies deposited in the target, the AMD and the RF-section are also shown in the table.

Table 1. Parameters and results using a drive electron beam of 6 GeV with beam size 4.0mm, hitting a target of 14mm thickness.

Parameters for target and captures		Parameters for the 300Hz scheme	
Drive beam energy	6 GeV	# drive e- /bunch	2×10^{10}
Beam size	4.0 mm (rms)	# bunches/triplet	132 (in 996 ns)
Target material	Tungsten	# bunches/train	2640 (in 63 ms)
Target thickness	14 mm	repetition of the trains	5 Hz
Max. AMD field	7 T	Results numbers in () are for the 300Hz scheme	
Taper parameter	60.1/mm	e+ yield	1.6 /e-
AMD length	214 mm	PEDD in the target	1.04 GeV/cm^3/e- (22.7 J/g)
Const. field	0.5 T	Energy deposit in the target	823 MeV/e- (35 kW)
Max. RF field	25 MV/m	Energy deposit in the AMD	780 MeV/e- (33 kW)
RF frequency	1.3 GHz	Energy deposit in the RF section	470 MeV/e- (20kW)

4. Conclusions

We propose the conceptual design of a conventional positron source driven by a several-GeV electron beam for ILC. A conventional scheme is the only experienced scheme of positron generation in real accelerators. By employing the experiences we can contain risk areas of positron generation in a target system. The drawback of a conventional positron source is that it has no capability to provide polarization. Replacement of the positron source is necessary, if we need a polarized positron beam in future. On the other hand in the viewpoint of a risk control, the conventional scheme has a significant advantage which is vitally important in designing a very large system such as ILC. The conventional positron source proposed in this paper largely reduces the risk in target system by pulse stretching, optimized beam parameter, and optimized target parameter. They cure target thermal issues and enable us to employ a conventional positron generation scheme in the ILC. From the viewpoint of risk control, the conventional scheme could be a suitable solution at the first stage of the ILC project.

Acknowledgements

We would like to acknowledge valuable discussions with Dr. M. Kuriki of Hirosima university. We wish to thank Dr. K. Yokoya of KEK, his critical comments were always useful to improve our ideas. We would like to thank fruitful discussions with Dr. L. Rinolfi of CERN and Dr. T. Kamitani of KEK. Our heartfelt appreciation goes to, Dr. S. Guiducci of INFN/Frascati gave us constructive comments on the relation between the damping ring design and positron source. We wish to thank Dr. J. Rochford of CCLRC/RAL and Dr. I. Bailey of Cockcroft Institute for their help to evaluate target heat issues. We also aknowleadge valuable discussion with Dr. R. Chehab of University Lyon-1 and Dr. A. Variola of LAL. A part of this research received support of Global COE Program "the Physical Sciences Frontier", MEXT, Japan.

References

1. ILC-Report-2007-001(2007),
   http://www.linearcollider.org/about/Publications/Reference-Design-Report.
2. V. Bharadwaj, et al., Proceedings of 2005 Particle Accelerator Conference, Knoxville, Tennessee, p. 3230.
3. G. Moortgat-Pick, et al., Physics Reports 460, 2008, pp131-243.
4. E. Reuter and J. Hodgson, PAC1991 Proceedings pp1996-9998, SLAC-PUB-5370.
5. T. Kamitani and L. Rinolfi, CLIC-NOTE-465, CERN-OPEN-2001-025.
6. S. Ecklund, SLAC-CN-128, 1981.
7. NLC Collaboration, SLAC-R-571, 2001.

MULTI BUNCH GMMA RAY GENERATION EXPERIMENT AT ATF*

TOHRU TAKAHASHI[†]

Graduate School of Advanced Sciences of Matter Hiroshima University, 1-3-1 Kagamiya Higashi-Hiroshima, 739-8530, Japan

We construct a new detector to monitor γ yields in bunch by bunch basis for the Laser Compton experiment at the KEK ATF which is capable to separate γ rays in 5.6ns-spacing multi-bunch operation of the KEK ATF. In this article we report a result of measurement of multi bunch γ ray detection for the first time at the KEK ATF.

1. Introduction

The polarized beams are powerful tool for physics at the International Linear Collider (ILC) since we can choose helicity Eigen states for the initial election-positron system. The ILC is designed to provide polarized electron beams using a DC gun with GaAs photo-cathode. In addition to the polarized electron, it is highly desirable to use polarized positron beams since it increase the effective polarization of the ILC[1], where effective polarization is defined as;

$$P_{eff} = \frac{P_e - P_p}{1 - P_e P_p}$$

(1)

For example, 80% polarization of the electron and 60% positron polarization yield the effective polarization of 95%. The ILC is planned to construct the positron source by the undulator scheme. In the undulator scheme, polarized γ rays of O(10MeV) are generated by feeding high energy electron beams in the main LINAC into a more than 150m long helical undulator, then γ rays are impinged into the metal target to generate positrons[2]. An optional scheme to make polarized γ rays is the Compton scheme. In Compton scheme, γ rays are

---

* This work is supported by "Grant-in-Aid for Creative Scientific Research of JSPS (KAKENHI 17GS0210)" project of the Ministry of Education, Science, Sports, Culture and Technology of Japan (MEXT) and Quantum Beam Technology Program of JST.

† Collaborators: T.Akagi, M.Kuriki, R.Tanaka, H.Yoshitama(Hiroshima University), S.Araki, Y.Funahashi, Y.Honda, T.Omori, T.Okugi, H.Shimizu, N.Terunuma, J.Urakawa (KEK, High Energy Accelerator Rearch Organization) H.Kataoka, T.Kon(Seikei University), K.Sakaue, M.Washio(Waseda University) and French-Japan Collaboration.

generated via Compton scattering of laser photons off electrons. The Compton scheme has several advantages such as; the electron energy necessary for O(10MeV) photons is O(1 GeV) which is much lower than those for the undulator scheme, the polarization of γ rays are easily controlled by the laser polarization. The principle of the Laser Compton scheme has been demonstrated by previous works [3]. While it is attractive for the polarized positron sources for the ILC, a price to pay for the Compton scheme is the intensity of the γ rays. We need high intensity and with high repetition laser pulses such as O(100mJ/pulse) with O(100MHz) repetition. We plan to achieve the required performance using a laser pulse stacking cavity in which laser pulses are accumulated coherently and its energy are enhanced inside the cavity. We have developed and tested prototype cavity at the KEK-ATF as was reported in [4].

One of the important issues for the ILC positron source is to generate multi-bunch positrons with stable intensity. The temporal bunch separation of the KEK-ATF is 5.6 ns and it is operative for 1 to 10 bunches per train, however, in the previous experiment, the γ ray detector equipped with CsI crystal did not have capability to observe γ rays from each bunch separately, because the decay constant of fluorescence is about 20ns even after discriminate faster component. Therefore we improved detection system to separate γ rays from laser-electron scattering which happen every 5.6 ns. In this article, we report preliminary results of multi-bunch photon detection from the laser Compton experiment for the first time at the KEK-ATF.

2. Experimental Setup

2.1. *Optical Resonant Cavity*

The schematic of the optical resonant cavity and laser optics are shown in Figure 1.

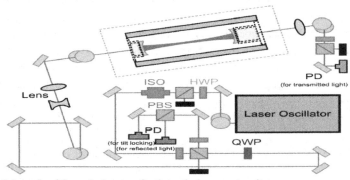

Figure 1. Schematic of the optical system for the optical resonant cavity.

We used a Mode Locked pulsed laser (COURGAR, Time-Bandwidth Products). The wave length of the laser is 1064 nm. The pulse energy is 25 nW with the repetition of 357 MHz. The optical resonant cavity is 2 miror Fabry-Perot type. The finesse of the cavity was measured to be 1800 which corresponds to the power enhancement factor of about 600. It was consistent with expectation from the reflectivity of mirrors of the cavity. A feedback system was constructed to keep the optical cavity on resonance with the laser pulse as well as to synchronize laser pulses with the electron bunches in the ATF. The schematic o the feedback system is shown in Figure 2.

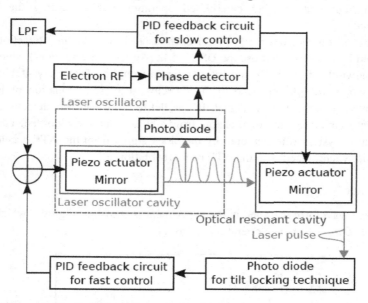

Figure 2.Diagram of the feedback and timing synchronization system.

The laser spot size inside the cavity at the laser-electron interaction point was about 30 μm (1 σ) which was estimated from the measured angular divergence the laser pulses outside the cavity. The laser power in the cavity was calculated from the measured transmitted power outside of the cavity. It was monitored and was about 1.5kW during the experiment.

2.2. Detector and Data Acquisition System

To detect γ rays in bunch by bunch basis, we installed new detector consisting of a BaF2 scintillator and a Photo-Multiplier Tube (PMT) R3377 (Hamamatsu

Photonics). Because the BaF2 crystal has two scintillation components with different wave length and decay constant, 310nm/600ns and 220nm/0.8ns, an optical filter was inserted between the crystal and the PMT to eliminate the slower component. The rise time of the PMT is 0.7 ns when operating at nominal high voltage. Prior to the beam experiment, detector was calibrated using cosmic rays. According to expected number of γ rays in the Compton experiment, the detector would be saturated if it was operated with the nominal voltage. Since the time response of the detector could be deteriorated for lower voltage, we took a cosmic ray data for pulse height as well as rise time with various high voltages and determined operation voltage during the beam experiment.

The waveform from the PMT for each laser electron beam interaction was recorded by a digital oscilloscope DSO5054a (Agilent Technologies) which has the bandwidth of 500MHz and the maximum sampling frequency 4GHz. The oscilloscope was placed close to the detector to avoid attenuation of higher frequency component in the signal. In parallel to the waveform, detector signal was sent to the ADC and total charge from a pulse train was recorded. The data acquisition system was triggered by the clock signal from the ATF accelerator which is synchronized with the ATF beam revolution.

3. Result

A typical waveform recorded in the experiment is shown in Figure 3 which was recorded during 10 bunch operation of the ATF.

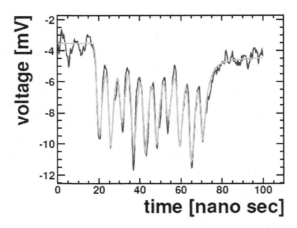

Figure 3. A typical waveform for 10 bunch data. The blue line is the raw detector signal. A result of a fitting taking the fluorescence and the detector response into account is overlaid by a green smooth line.

We clearly observed spikes of 5.6 ns separation in the waveform which correspond to γ rays from each bunch in a train. A broad shift of the baseline can be understood as a contribution of leakage of the slower component of fluorescence from the optical filter.

To estimate number of photons we fitted the waveform with a function which takes into account time dependence of the fluorescence in the BaF2 crystal, light correction in the crystal and detector response. A result of fitting is superimposed on the waveform in Figure 3. The beam current for each train in the ATF was measured the DC Current Transformer (DCCT) while relative intensity of between electron bunches was monitored by the Wall Current Monitor (WCM).

The number of γ rays for bunch by bunch basis normalized the laser power and the electron intensity is shown in Figure 4.

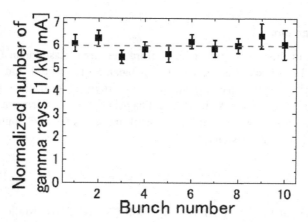

Figure 4. Normalized γ ray yields from each bunch for the 10 bunch operation. The dashed line is an expectation from numerical calculation.

The expected γ ray yields was estimated by CAIN [5] program with the laser and the electron beam parameters. The γ ray yields were consistent with the expectation within the error and no bunch dependence for γ ray yield was observed.

4. Summary

The detector system of the laser Compton experiment at the KEK-ATF was improved to separate γ rays from each electron bunch in multi-bunch operation. The γ ray yields were observed with bunch by bunch basis up to 10 bunches/train. Observed γ ray yields were consistent with numerical expectation and no bunch dependence of the yield was observed. It was an encouraging step for the positron source R&D. Several works are in progress to improve performance of the detector and the data acquisition system toward future experiments;

- New optical filter will be inserted between the BaF2 crystal and the photo cathode of the PMT to cut leakage of slower component from BaF2 crystal.
- The signal output circuit of the PMT will be modified to avoid saturation for more intense γ ray yield which will be expected in future experiment.
- New data acquisition system is under development to record the data for every beam revolution in the ATF.

Acknowledgments

Data shown in this article was taken before the Great East Japan Earthquake. The KEK-ATF is back after the Great East Japan Earthquake in June 2011 after the earthquake and precise alignment is in progress for full recovery. The authors would like to thank all ILC colleagues for their warm messages and supports. Particularly we would like to thank members for ILC source working group for valuable discussions.

References

1. G. Moortgat-Pick et al., Phys. Rep 460 (2008) 131.H. Müller and B. D. Serot, Phys. Rev. C52, 2072 (1995).
2. ILC-REPORT-2007-001(2007)
3. M. Fukuda et al., Phys. Rev. Letts 91 (2003) 164801.T. Omori et al., Phys. Rev. Letts 96 (2006) 114801.
4. For comprehensive summary of previous works, see S. Miyoshi, Ph D. thesis,
5. ADSM Hiroshima University, 2010.
6. http://www.huhep.org/Home/thesis/doctor/2011miyoshi.pdf
7. http://lcdev.kek.jp/~yokoya/CAIN/cain235/

POSITRON SOURCE USING CHANNELING WITH A GRANULAR CONVERTER

P. SIEVERS[1], C. XU[2,3], R. CHEHAB[4], X. ARTRU[4], M. CHEVALLIER[4],
O. DADOUN[2], G. PEI[3], V.M. STRAKHOVENKO[5], A. VARIOLA[2]

[1]CERN,[2]LAL, [3]IHEP, [4]IPNL, [5]BINP

A new kind of positron source is presented. It uses the intense radiation emitted by channeled electrons in axially oriented crystals to generate positrons in a granular amorphous converter. A distance of 2 meters between the crystal radiator and the amorphous converter is used to put a bending magnet which sweeps off the charged particles allowing only to the photons to hit the converter. The simulation results on photon production, and positron generation are reported. A particular attention is brought to the energy deposition and Peak Energy Deposition Density (PEDD) in the targets. Emphasis is put on the problems of heating, cooling and on different methods and devices to handle the power dissipation in the targets. An application for the case of ILC (International Linear Collider) is worked out.

1. Introduction

Following theoretical investigations on radiation in oriented crystals by ultrarelativistic particles [1], simulations have been carried out to investigate the possibility to use this intense radiation to generate intense positron beams [2]. Experiments worked out at LAL-Orsay, Tokyo University, CERN and KEK [3,4,5,6,7,8] confirmed the interest of this method and provided useful informations on the positron phase space [9]. In the framework of the studies on the future linear colliders, in which the electron driver beam has intensities of about 10^{14}e-/s, a particular attention was brought to the energy deposition density and power deposition in the targets. Our investigation is aiming to address these problems proposing solutions which allow an easier thermal dissipation and henceforth, a better lifetime for the target. We have considered a particular shape for the amorphous converter which is made of a great numbers of small W spheres. Such solution was previously considered for the neutrino factories were multi-megawatt proton beams are impinging on the targets [10]. We present, hereafter, the main results concerning the photons and the positrons for a hybrid source made of a crystal radiator and a granular amorphous converter placed 2 meters downstream with a sweeping bending magnet in between.

2. The Hybrid Source

A high energy electron beam (5 to 10 GeV) is directed along the crystal axis. Photons, electrons and positrons are generated. A sweeping magnet takes off the charged particles and only the photons impinge on the amorphous converter (figure 1). The distance may be about 2 meters. For the ILC, we consider an electron beam of 10 GeV with a transverse rms radius of 2.5 mm. A scheme of the hybrid source is presented on figure 1.

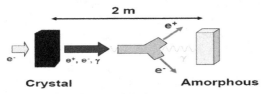

Fig. 1. The hybrid positron source

2.1. *The crystal radiator*

The Positron Source relies upon an intense beam of high-energy photons impinging upon a metal target. The photons are generated by radiation from relativistic electrons as they pass through a crystal tungsten radiator. The <111> crystal axis is aligned on the direction of the 10 GeV impinging electron beam. Due to channeling radiation an intense photon beam with an average energy of 300 MeV emerges at the exit of the radiator. The radiator thickness has been chosen 1 mm, it corresponds to an optimum value. The electrons being channeled along the axis exhibit an oscillating trajectory with a period of ~ 1μm corresponding to several interatomic distances. The frequency of the radiated photons is roughly given by:

$$\Omega = 2\gamma^2 \, \Delta E_T \tag{1}$$

Where γ is the relativistic Lorentz factor, ΔE_T is the energy gap between two channeled states.

2.2. *The granular converter*

The very intense incident electron beam considered in the linear collider projects (10^{14} e-/second) requires resistant targets. In the muon collider and neutrino factories projects, a similar situation is met with powerful proton beams impinging on targets to produce pions decaying into muons and then into neutrinos. P.Sievers and P. Pugnat have proposed the use of granular targets made of a great quantity of small spheres [10]. The high ratio of surface/volume of the spheres (~ 3/r) makes easier the thermal dissipation. In the scheme considered for the converter of the hybrid source, the spheres are arranged in staggered rows. An example is shown on figure 2.

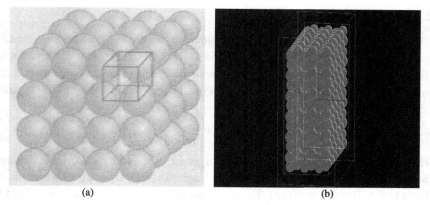

(a)   (b)

Fig. 2. A Granular target : (a) 4 rows of r = 1 mm spheres; (b) staggered rows

The choice of the diameter of the spheres and of the number of rows is related to two quantities:

- The required positron yield
- The Peak Energy Deposition Density (PEDD)

A comparison between two granular targets and a compact one giving close yield values is presented on Table 1.

Table 1 - Comparison of granular and compact positron targets

	Thickness (mm)	Yield (e+/e-)	PEDD (GeV/cm^3/e-)	ΔE MeV/e-	N-Layer	Spheres number	Effective density(g/cm^3)
Compact	8	13.3	2.24	523			19.3
Granular r=1mm	10.16	12.5	1.8	446	3	864	13.9
Granular r=0.5mm	11.60	13.45	2.33	613	7	8064	13.9

The simulations have been worked out using two programs :

- A program describing the crystal effects (channeling radiation, coherent bremsstrahlung, pair creation,..) from Prof. Strakhovenko and referred as VMS [11].
- The GEANT4 program using the outputs of the first program as event generator; the results on positron yield, energy deposited and PEDD in the granular target are provided by GEANT4.

Considering the yield values, the energy deposition per incident electron (on the crystal) and the PEDD value, we have chosen the 3-layer (having 6 rows)

120

target; its comparable low PEDD value with respect to the others looks interesting. On the other hand, spheres with 2 mm diameter seem easy to obtain.

2.3. The photons

The photons coming out from the crystal are impinging on the amorphous granular target. Its transverse dimension is about 3.5 mm rms radius. Such dimension is small enough for the geometrical acceptance of the matching system put after the target while still achieving reasonable PEDD in the converter.

2.4. The positrons

The positrons generated in the granular target are captured by a special matching device AMD: an adiabatic lens with a magnetic field tapering down from 6 Tesla to 0.5 Tesla over 50 cm. The positrons are, then, accelerated with a field | E |= 15 MV/m. The positron yield after capture and acceleration over one meter structure is about 2.8 e+/e-.

2.5. Energy deposited and PEDD

The configuration chosen for the linear collider ILC is based on a pulse time structure modification before the target in order to decrease the power deposited per pulse; the nominal structure of ILC beam is then recuperated after the Damping Ring. For that purpose, we have chosen the configuration proposed by T.Omori from KEK [12] and where the incident electron beam is made of minitrains of 100 bunches each with a periodicity of 300 Hz in a macropulse of 40 ms containing 13 minitrains. The scheme of the 300 Hz solution as defined by Omori et al. is presented on Figure 3.

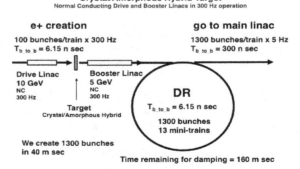

Fig. 3. The scheme of the 300 Hz solution proposed by Omori et al.

3. Heating and cooling of the targets

3.1. *The crystal*

In the following, the design of crystal targets and their cooling systems are discussed. The following input parameters and data, as computed with the above cited codes have been used. The energy of the electron beam, incident on the crystal is 10 GeV, and the beam width, 2.5 mm rms. Concerning the dimensions of the tungsten crystal, it is a disk with a diameter of 1 cm and a thickness of 1 mm. The maximum PEDD, located on axis at the downstream face of the crystal is 5.83 J/g per micro pulse and the total energy deposited in the crystal is 2.6 J/micro pulse.

3.1.1. *Cooling of the crystal*

In Fig.4 two lay outs for the crystal units and their cooling is shown.
(a) : the deposited heat is evacuated radially through the crystal by cooling its edge which is clamped or brazed to a water cooled frame.
(b) : the deposited heat is evacuated through the two side faces of the crystal, cooled by forced convection via a Helium gas stream .This, clearly, requires the containment of the He gas by windows on either side of the crystal.
A scheme of the crystal holders is given on Figure 4.

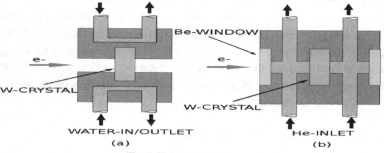

Fig. 4. The crystal holder

The instantaneous temperature rise at the hottest spot of the crystal will be 41 °K/micro pulse. Now, in a completely adiabatic regime, when during a macro pulse of 40 ms duration, heat diffusion within the crystal and cooling are ignored, a maximum, local temperature rise of 533 °K/macro pulse would be expected at the "hottest" spot. However, taking into account the above effects, smaller temperature jumps per macro pulse, averaged over the crystal of 242 °K and 176 °K are estimated for the water and He-cooling respectively. This demonstrates the higher efficiency of He-cooling, taking place close to the "hottest" spot, relative to the water cooling, located at the more distant edge of the crystal.

This effect, however, is reduced due to the relatively low mass flow of He (He at 10 Bar, a velocity of 20 m/s and a mass flow of 0.72 g/s), which leads to elevated temperatures in the He, rising by 210° K/macro pulse when passing the crystal. This temperature offset has to be added to the temperature in the crystal. For water, this effect is much lower, 13 °K/macro pulse.

Finally we arrive at maximum temperatures in the crystal after a macro pulse of 275 °C and 406 °C for water and He-cooling respectively.

3.1.2. *Thermal stress in the crystal*

Considering now the thermal stress, induced in the crystal by these thermal cycles of 41 °K/micro pulse, stresses of about 75 MPa (about 5% of the elastic limit of solid Tungsten), will be created, independent from the cooling method. With about 561 Million. pulses per 100 day of continuous operation, material fatigue may become an issue, even at these low stress values, in particular for oriented crystals.

The stresses in the He-cooled crystal during a macro pulse, where only little thermal gradient prevail, can be made very small. This can be achieved by mounting it in a frame, allowing for radial and axial thermal expansion, still maintaining of course a precise orientation in angle (0.5 mrad).

For the edge cooled crystal, mounted into a rigid frame with good thermal contact, stress levels of 468 MPA/macro pulse (30% of the elastic limit, 43 Million. pulses/100 days) may result. This can however be mitigated by using elastic materials for the frame, like Aluminum or Titanium, which will allow for some radial thermal expansion of the crystal and thus reduction of its stress. This, certainly, requires more studies.

3.1.3. *The beryllium windows of the crystal unit*

Windows for the He-cooled crystal are required, to contain the cooling fluid, while this is not the case for the water cooled option. These windows will have a free diameter of about 1 cm and a thickness of 1 mm, to resist to the He-pressure of 10 Bar. The maximum beam effects occur, obviously, in the downstream window. Similarly to the W-crystal, the Be-window will also be submitted to thermal, cyclic stress. Moreover, care must also be taken for the mounting of this window into an elastic frame (Al, Ti), to allow for its radial thermal expansion in order to reduce the thermal stress induced by each macro pulse.

3.1.4. *An array of five crystals to extend the life time of the unit*

One may consider mounting five, precisely aligned crystals, in a row into the goniometer. With a moderate velocity of displacement of this array of 5 cm/s, each crystal will be hit by 13 micro pulses per 40 ms, as for the stationary crystal, but in the average by only one macro pulse/s. However, the crystals placed at the end of the array will be hit every so often twice within 0.2 s.

Since now there is, in the average, more time available between pulses incident on each of the five crystals, the average power and absorbed dose per crystal will be lower by a factor of five, as compared to a single, stationary crystal. However, short term effects, like peak temperatures after each macro pulse in the W-crystal will only change moderately: 178 °C and 365 °C for water and He-cooling respectively. Similar arguments hold for the Be-windows of the He-cooled target array. A peak temperature of 76 °C was estimated.

The remaining thermal stress in the window requires still a careful design of the frame (see above). However, the reduced number of pulses per crystal, and thus fatigue effects (112 Million micro pulses/100 day and 8.6 Million. macro pulses/100 day) should lead also to five times longer life times.

3.2. *The granular amorphous target*

The amorphous W-target for elevated positron production requires a thick target with a volume of about 1 cm^3. This leads to very high PEDD's and average powers, difficult to cool.

Therefore a granular target is proposed which consists of an ensemble of W-spheres with a radius r of 1 mm. Helium is used as a cooling fluid, passing through the voids between the spheres. This provides a very efficient cooling applied to each sphere, due to its high surface to volume ratio of 3/r, in particular for small spheres.

Furthermore, since the more or less constant temperature gradients across each sphere are small, very low thermal stresses will occur within each sphere. Also, small spheres will suffer much less from thermal shock (see later) than bulk material.

Effective densities of 75% of solid W can be achieved by the densest possible packing. In the following, however, we take a more conservative density of 72 %, as used in the simulations.

For the cooling of the spheres by forced convection, with a He-stream at a pressure of 10 Bar and an entrance velocity of 10-20 m/s, a convection coefficient of about 1 W/cm^2 and °K can be expected, close to values achieved with water cooling. He is chosen for its chemical inertia and compatibility with

W at elevated temperatures, while water will lead to corrosion of W. One may also consider as an option, spheres made from heavy, noble metals, like Platinum, compatible with water cooling. However, water hammer induced by short beam pulses will become an issue, Thus, water cooling is not considered further at this stage.

In the following we assume a granular target with a cross section of 1 x 1 cm^2 and a thickness of 1 cm. It is hit by the γ-beam, emerging from the upstream crystal at an average energy of 300 MeV and an rms. width of 3.5mm. A maximum PEDD of 31.5 J/g and micro pulse is reached on the axis at the downstream end of the target. The average energy, deposited in the whole target volume is 146 J/micro pulse. This leads to a temperature rise of 222 °K/micro pulse at the "hottest spot" in the target. Per macro pulse these values will be 13 times higher since little thermal diffusion through the granules and little cooling occurs during a macro pulse of 40 ms duration. This clearly demonstrates that a stationary target will not sustain the thermal loads and moving targets have to be considered.

3.2.1. *The moving pendulum target*

In order to prevent pile up of two subsequent micro pulses, the array of 13 targets must be displaced with a velocity of about 3 m/s. Arranging the targets around the rim of a rotating wheel, would require a diameter of about 4 cm, rotating at 1400 rpm. In view of these rather high rotation frequencies and loads on the rotating seals, required to inject and recuperate the cooling fluid, trolling or wobbling targets have been considered (See the presentations of the Durham ILC e+ meeting, October 2009.). In the following we describe a "Pendulum Target" as sketched in Fig. 5. An array of 13 targets is arranged on a circular sector of 13 cm width. The pendular displacement is transmitted through a pivoting shaft with a length of about 50 cm from the outside through a vacuum tight bellows. Simultaneously, the cooling fluid from the outside is injected and recuperated through this shaft, thus avoiding rotating seals. Clearly, a flexible connection for the fluid at the outside top of the shaft must be provided.

Transmitting a sinusoidal movement to the shaft (not detailed in the picture), the unit must pivot through a total angle of +/- 24 degree (A and C), of which the central range of ± 7 deg., where the velocity is about constant, is used for the beam impacts. At the extremes, sufficient range in angle of 17 deg., corresponding to 160 ms, remains during the beam off time for the smooth inversion of the movement of the pendulum. A total cycle of the pendular movement takes 400 ms, i.e. 2.5 Hz (22 Million cycles/100 day), also to be sustained by the bellows. In the average, each target will be hit every 200 ms, where of course, the extreme targets will be hit twice every so often within 160 ms or 240 ms.

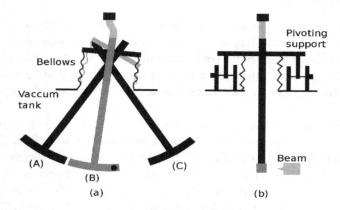

Fig. 5. The pendulum for a granular target

3.2.2. *Cooling of the pendulum*

As stated before, 146 J/micro pulse are deposited, leading to an average power of 730 W/target. The total power, to be removed from the pendulum, is 9.5 kW. The peak temperature at the "hottest sphere" will be 248 °K/micro pulse, to which the temperature rise of the Helium, passing it, of 75 °K must still be added. Finally, the hottest sphere may well reach 420° C. For water cooling, about 250° C would result.

Considering now the Be-windows, required for this He-cooled pendulum, the PEDD there is 30.6 J/g and micro pulse and the total deposited energy is 3.4 J/micro pulse. This leads to a maximum, adiabatic temperature rise in the Be of 16.3 °K/micro pulse and to an average power in the window of 17 W in each of the 13 target segments.

Collecting now the different temperature rises in the window and the passing He, a final, maximum temperature of about 186° C in the Be has been estimated. As above, care must be taken to separate the hotter He-stream through the W-granules from the cooler one passing the Be-window. Also, the thermal stresses in the window

(43 Million pulses/100 day) cannot be neglected. Care must be taken to mount the window, which is rather a circular strip covering the array of targets, into a frame with sufficient elasticity, to allow for thermal expansion of the Be.

3.2.3. *The rotating wheel*

Although it is of interest to avoid rotating seals, as described above for the pendulum, high oscillation frequencies and elevated temperatures in the W-granules and the Be-window of the pendulum will result. Thus, it is of interest to use a true rotating wheel, fitted with more than 13 "targets" and using rotating seals for the cooling fluid as well as for the vacuum. A sketch of the rotating wheel is given in Fig.6.

As above, a linear velocity of about 3 m/s of the rim of the wheel is required to avoid pile up of subsequent micro pulses. In particular, we consider a wheel with a diameter of 58 cm, rotating at 107 rpm and fitted with 182 "targets" of 1 cm diameter. The average time between two hits of the same target of 2.8 s is much longer than for the pendulum. This provides more time for cooling and reduces the temperatures due to the lower average power in each target to 52 W. Still the total power, to be removed from the wheel remains, as before, 9.5 kW.

Applying of course the same thermal loads per micro pulse, but with a time slot of 2.8 s between two hits of the same target, no pile up occurs in the targets, each will be cooled down to room temperature before being hit again. This leads to maximum temperatures of 242° C and 36° C in the W-granules and Be-window respectively. Moreover, the material fatigue will be considerably be reduced with now 3 Million pulses/100 days for each target area. A sketch of the rotating wheel is given on Figure 6.

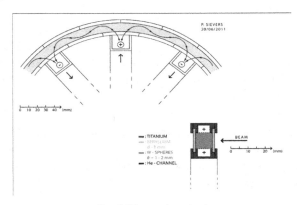

Fig. 6. The rotating wheel

3.2.4. *Thermal shock*

The very short micro pulses with a duration of $t_0 = 0.6$ μs give rise to thermal shocks in the crystal as well as in the amorphous target. The resulting instantaneous, adiabatic temperature rise ΔT over 0.6 μs in the Tungsten material are estimated as 41 °K and 222 °K in the crystal and the target respectively. The normally expected thermal expansion will not take place over the very short heating up time t_0, since any displacement of the irradiated volume inside the target is prevented by its mass inertia. Thereafter, of course, pressure waves will start to propagate from this instantaneously heated zone. In the case of small spheres, this situation is equivalent to a statically compressed sphere, where at time zero the retaining pressure at the boundary is instantaneously removed: a relaxation wave will propagate from this boundary towards its centre.

The maximum initial pressure σ for instantaneous heating of a sphere is given by

$$\sigma \cong 3 \, E \, \alpha \, \Delta T, \tag{2}$$

E: Young's Modulus; α: linear thermal expansion coefficient.

For thin disks which can rapidly relax into axial direction, the stresses will be about 50% lower. As a "relaxation time" for the pressure one may define the time $t(tr)$ it takes the pressure wave, propagating with the velocity c of sound (in W: $c \cong 4$ km/s), to travel from the boundary of the heated zone to its center

Considering now the W-crystal, where the radius of the mostly heated area is about 2.5 mm, the time $t(tr)$ is about 0.6 μs, close to the heating up time to. Thus, only little relaxation of the shock may be expected and one may have to account for a maximum stress σ of 112 MPa (about 7% of the yield strength of W). Following a similar argument for the downstream Be-window of the crystal unit, a pressure of 6 MPa (15% of the yield strength of Be) will result.

Applying now the same arguments for the amorphous target with its small W-spheres with a radius of 1mm, the transition time $t(tr)$ will be of the order of 0.25 μs, significantly below the heating up time to of 0.6 μs. Thus, the stress due to each micro pulse in the small spheres will be below the maximum stress σ with instantaneous heating. Stresses of 500 MPa/micro pulse (33% of the yield strength) in the W-spheres and 46 MPa/micro pulse (20-50% of the yield strength) in the downstream Be window are estimated.

Table 2 - Summarizing table Brackets indicate the cooling case

	1 Crystal	5 Crystal	Pendulum		Rot. Whell	
	W	W	W	Be	W	Be
PEDD (J/g)	5.83	5.83	31.5	30.6	31.5	30.6
ΔT/micro (K)	41	41	222	16.3	222	16.3
En/micro p. (J)	2.6	2.6	146	3.4	146	3.4
Av Power (W/Targ.)	170	34	730	17	52	0.24
Max. Temp °C,[He]	406	365	420	186	242	36
Max. Temp °C,[Water]	275	178	250	-	-	-

The above should, however, be considered only as rough estimate. More sophisticated computations will be required. But even so, more subtle effects, like the crystal structure in the material and its evolution under cyclic heating and irradiation, might lead to degradations, very difficult to assess. Therefore, prototyping, including tests with beam, will be essential.

A summarizing table for the heating and cooling data is presented on table 2. Some calculations on Be windows of the crystal holder are under development.

3.2.5. *Eddy Currents*

The electromagnetic forces, induced in rotating wheels, immersed partially in the pulsed magnetic field of the downstream AMD structure, have been studied for fast rotating Titanium wheels (500 rpm) with a linear velocity of its rim of about 30 m/s (Durham ILC e+Meeting, Oct. 2009). With the velocity of the wheel considered here of only 3 m/s, such forces will be considerably reduced. The induced current density, interacting with the outside magnetic field, depend also on the electrical conductivity of the rotating components. This should be lower in the ensemble of the W granules than in bulk material. In the continuous Be-window, however, the induced currents must be considered more carefully.

4. Summary and conclusions

The hybrid target with a granular amorphous converter exhibits:
- a good quality positron beam with high yield generated by photons from channeling radiation. The yield value is corresponding to the requirements of the future linear colliders.
- a PEDD for the granular converter is lower than for the compact one allowing a better lifetime

- an efficient thermal dissipation process due to the granular character.
- The pulsing of the incident beam creates, however, cyclic loads in temperature and thermal stress in the target materials. Effects, like material fatigue in combination with radiation damage are difficult to assess. Therefore, to increase the lifetime of the target, the incident beam power must be spread cross moving targets or rotating wheels. Clearly, experimental verifications and tests will be required to verify predictions.

Acknowledgements

The authors are indebted to T.Suwada, T.Kamitani, T.Omori, J.Urakawa from KEK, T.Takahashi from Hiroshima University and L.Rinolfi and A.Vivoli from CERN for valuable discussions

References

1. V.N. Baier, V.M. Katkov, V.M. Strakhovenko "Radiation yield of high-energy electrons in thick crystals" Phys.Stat.Sol. (b) **133**, (1986) 583-592
2. X.Artru, V.N.Baier, R.Chehab, A.Jejcic "Positron source using channeling in a tungsten crystal" Nucl.Instrum.Methods A **344** (1994) 443-454
3. X.Artru et al. "Axial channeling of relativistic electrons in crystals as a source for positron production" Nucl.Instrum.Methods B **119** (1996) 246-252
4. X.Artru, R.Chehab, M.Chevallier, V.M.Strakhovenko "Advantages of axially aligned crystal in positron production at future linear collider" Phys. Rev. Special Topics A-B **6** (2003) 091003
5. R.Chehab et al. "Experimental study of a crystal positron source" Physics Letters B **525** (2002) 41-48
6. X.Artru et al. "experiment with a crystal-assisted positron source using 6 and 10 GeV electrons" Nucl.Instrum.Methods B **201** (2003) 243-252
7. T.Suwada et al. "Measurement of positron production efficiency from a tungsten monocrystal using 4 and 8 GeV electrons" Phys.Rev. **E 67** (2003) 016502
8. T.Suwada et al. "First application of a tungsten single crystal positron source at the KEKB factory" Phys.Rev. Special Topics A-B **10** (2007) 073501
9. X.Artru et al. "Summary of experimental studies at CERN on a positron source using crystal effects" Nucl.Instrum.Methods B **240** (2005) 762-
10. P.Sievers, P.Pugnat "a He-gas cooled stationary target" J.Phys. G: Part.Phys. **29** (2003)
11. V.M.Strakhovenko (programme VMS)
12. T.Omori (see POSIPOL 2009 and POSIPOL 2010)

COMPTON e⁺ SOURCE OVERVIEW (RING + ERL +LINAC)*

JUNJI URAKAWA†

KEK, Accelerator, 1-1 Oho
Tsukuba, Ibaraki 305-0801, Japan

Short reviews of Compton e⁺ sources R&D for ILC are given. Especially, the Compton sources using Ring and ERL are discussed according to the scheme of truly 300Hz conventional e+ source with the change of the positron generation scheme from non-polarized to highly polarized e⁺. As we have proposed the Ring Compton source since 2005, the improvement of the design is described. The technologies for X-ray and Gamma-ray source based on the inverse-Compton scattering have been improved by recent fiber laser and high finesse optical cavity developments.

1. Short Reviews of Compton e⁺ Sources R&D for ILC

There are three schemes for polarized positron beam generation based on the inverse-Compton scattering. They are called Ring Compton, ERL Compton and Linac Compton. The Ring Compton consists of 1.8GeV Linac, a storage ring with five Compton interaction points (IPs) at which an optical cavity is installed to keep laser pulse of 600mJ in it, conversion target, capture system and 5GeV booster Linac to inject the positron beam into damping ring [1, 2].

ERL Compton; ERL (Energy Recovery Linac) is employed as the dedicated electron driver which accelerates 480pC bunched beam up to 1.8GeV with bunch spacing of 18.5ns. Then, multi-bunch beam collides the laser pulse of 600mJ at five IPs at which an optical cavity is installed to keep 600mJ laser pulse [3].

Linac Compton; Conventional Positron source (Electron Linac driven one) is upgraded with a simple modification. Polarized γ-ray beam is generated in the

---

* This work is supported by the Quantum Beam Technology Program of MEXT.

† Collaborators: S. Araki, Y. Honda, T. Omori, T. Okugi, H. Shimizu, N. Terunuma, H. Hayano, K. Kubo, K. Watanabe (KEK, High Energy Accelerator Research Organization), T. Akagi, M. Kuriki, R. Tanaka, H. Yoshitama, T. Takahashi, H. Iijima (Hiroshima University), K. Sakaue, M. Washio (Waseda University), H. Kataoka, T. Kon (Seikei University), E. Bulyak, P. Gladkikh (NSC KIPT, Kharkov, Ukraine), F. Zimmermann (CERN), French-Japan and PosiPol Collaboration

Compton back scattering inside optical cavities of CO_2 laser beam with 6GeV electron beam produced by the Linac. The required intensities of polarized positrons are obtained due to 5 times increase of the e-beam charge (compared to non polarized case) and ten CO_2 laser system IPs [4, 5].

2. Change of the Positron Generation Scheme from Non-Polarized to High Polarized e⁺

Since 300Hz truly conventional non-polarized positron source was proposed recently, three polarized Compton source should be matched to the 300Hz normal Linac for smooth change from non-polarized to high polarized positron source [6]. Linac Compton will be smoothly replaced if the reliability of the CO_2 laser technology and the heavy beam loading compensation scheme will be confirmed with about 60% polarization [7, 8].

Especially, the Compton sources using Ring and ERL are discussed according to truly 300Hz conventional e+ source with the change of the positron generation scheme from non-polarized to high polarized e⁺. The heavy beam loading Linac is usually operated within about 1μsec pulse duration due to normal conducting energy consumption [9]. So, we have to assume about 1GeV stacking ring with about 260m circumference. Then, we can reuse 5GeV heavy beam loading Linac to inject bunch train of 3×10^{10} positrons/bunch into the damping ring within 63ms. Regarding Ring Compton since the period of 100 times positron bunch stacking is about 87μs, the cooling period of 3.2ms for Compton ring and the stacking ring is enough for stable operation of the ring. 20 times beam extraction from the stacking ring means it takes 63ms by 300Hz Linac operation. Of course, we have to construct 5Hz superconducting Linac with 63ms pulse duration and about 1GeV acceleration. Regarding ERL Compton since we need 1000 times of the beam stacking, the period of 1000 times positron bunch stacking is about 2.5ms for ERL Compton. About 1GeV superconducting Linac with 63ms pulse duration has 20 trains which consist of 132x1000 bunches train in 2.5ms with train spacing of 0.8ms for the bunch train extraction. The 0.8ms for the stacking ring is enough for the stable beam extraction operation. Then, we can reuse 300Hz Linac to accelerate and to inject into the damping ring within 63ms [10].

3. Design of the Compton Source

Table 1 shows the design parameters for Ring Compton, ERL Compton and Linac Compton. Bunch charge, bunch spacing and number of IPs strongly depend on ILC design requirements and accelerator technologies [11]. Also, we are assuming the improvements of the laser technologies in these design parameters in the future. Recently we proposed the head-on collision scheme to increase the yield of gamma-ray due to the inverse Compton scattering by the factor of ~3 and the decrease of the laser energy in the optical cavity by the factor of ~10 because of the damage of the high reflective mirrors [12].

As we have proposed the Ring Compton source since 2005, the improvement of the design is described here [13]. Our papers initiated the possibility to make the positron beam with the high polarization more than 80% based on the inverse Compton scattering [14, 15, 16]. In the report of our proposal we assumed 30 IPs to generate the necessary gamma yields but the design of the Compton ring and the laser system was upgraded such as the number of IPs is reduced to 5. The important improvements came from the study of the beam dynamics in the ring with the intense laser pulse and the continuous R&D for the optical cavity to generate the high brightness gamma-ray and X-ray [17].

Table 1. Design parameters for Ring Compton, ERL Compton and Linac Compton.

	Ring Compton	ERL Compton	Linac Compton
Bunch charge	5.3nC	0.48nC	10nC
Energy	1.8GeV	1.8GeV	6GeV
Bunch spacing	6.15ns	18.5ns	12.4ns
Bunch length at IPs	1ps	1ps	5ps
Laser energy at IPs	600mJ-1μm laser	600mJ-1μm laser	5J-10.6μm CO_2 laser
Number of IPs	5	5	10

4. Technologies for X-ray and Gamma-ray Source Based on Inverse-Compton Scattering

The technologies for X-ray and Gamma-ray source based on the inverse-Compton scattering have been improved by recent fiber laser and high finesse optical cavity [18 - 22]. We are going to demonstrate the generation of high brightness Gamma-ray at ATF damping ring and upgrading the linear accelerator technologies to accelerate high intensity electron beam for the generation of the high brightness X-ray. Two sets of 3D four mirror optical cavities have been installed in the damping ring and high brightness gamma-ray has been measured [23]. In this experiment, we established the technology of the high finesse optical cavity with about 5000, which has the laser waist of about 15μm in σ. At present the measured yields of gamma-ray do not satisfy the requirement of ILC by factor of ~20000 [24]. We have to develop the technologies to enhance the laser pulse energy by the factor of ~1000 and the intensity of electron bunched beam by ~20. Regarding the development of the fiber laser system, 100mW mode-locked laser oscillator, the fiber amplification to get 50W average power of 162.5MHz mode-locked laser and burst amplification by 70 times are established. Therefore, we can produce 21μJ/pulse and will demonstrate the accumulation of the laser pulse with 600mJ in the 4-mirror optical cavity in the future. Of course, we have to take care of the mirror coating damage due to high power laser by enlarging the size of laser on the mirror. Fig. 1 shows the layout of the ATF for the generation of the high brightness gamma-ray.

On the other hand, KEK is constructing the superconducting linear accelerator for Quantum Beam Technology Program (QBTP) under the support of MEXT. The purpose of this program is to demonstrate the generation of the high brightness X-ray in small accelerator for the application on Life, Chemistry, Bio and Environment sciences. The target values for the photon generation is 5 x 10^{10} photons/pulse/10% bandwidth. Fig. 2 shows the photo of QBTP construction at STF (Superconducting accelerator Test Facility) and the layout of the experimental system for the X-ray generation experiment. In this autumn 2012, the X-ray detector will measure the high flux to confirm the scheme of Compton technologies.

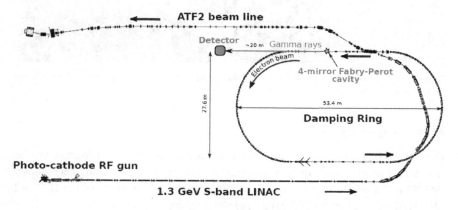

Figure 1. The Accelerator Test Facility (ATF) at KEK. This shows the location of 4-mirror Fabry-Perot cavity and gamma detector which are used for the generation of the polarized gamma-ray. This draw adapted from [25,26,27].

134

5. Summary

We succeeded the reduction of the number of IPs from 30 to 5 for ILC polarized positron beam generation based on the Compton scheme. However, we still need more effort to confirm the reliable technologies on the laser system with the optical cavities and the high intensity accelerator system.

Figure 2. The photo of QBTP construction and the layout of the experimental system for the X-ray generation experiment.

Acknowledgments

This work is supported by the Quantum Beam Technology Program of the Japanese Ministry of Education, Culture, Sports, Science, and Technology (MEXT). We would like to express our gratitude to all members of the ILC and ERL R&D's group at KEK for their helpful support.

References

1. J. Urakawa, Polarized e+ source for ILC based on Compton scheme, presentation in PosiPol 2006 Workshop at CERN.

2. M. Kuriki et al., ILC positron source based on laser Compton. AIP Conf. Proc., 980:92–101, 2008.
3. A. Variola, A Compton based polarized positron source for the SuperB factory. First considerations, presentation in PosiPol 2007 Workshop at LAL.
4. V. Yakimenko, Polarised e^+ source for ILC based on Compton backscattering of high intensity of electron and CO_2 laser beams, presentation in PosiPol 2006 Workshop at CERN.
5. I. Pogorelsky, Review CO_2 laser system for ILC polarized gamma source based on intra-cavity Compton backscattering, presentation in PosiPol 2006 Workshop at CERN.
6. T. Omori et al., Nuclear Instruments and Methods in Physics Research A **672**, 52–56(2012).
7. I. Pogorelsky, Polarized gamma-source based on intra-cavity Compton backscattering, presentation in PosiPol 2009 Workshop at Lyon.
8. M. Babzien et al., Observation of the Second Harmonic in Thomson Scattering from Relativistic Electrons, Phys. Rev. Lett. **96**,054802(**2006**).
9. S. Kashiwagi et al., Beam Loading Compensation using Phase to Amplitude Modukation Method in ATF, LINAC98 international conference proceedings, 91-93, 1998.
10. M. Kuriki, Compton based e+ sources for ILC/CLIC, presentation in PosiPol 2010 Workshop at KEK.
11. ILC-Report-2007-001(2007);/http://www.linearcollider.org/about/ Publications/Reference-Design-Report.
12. J. Urakawa, Development of a compact x-ray source based on Compton scattering using a 1.3 GHz superconducting RF accelerating Linac and a new laser storage cavity. NIM A, 637(1, Supplement 1):S47 – S50, 2011.
13. S. Araki et al., Design of a polarized positron source based on laser Compton scattering. Arxiv preprint physics/0509016, 2005.
14. I. Sakai et al., Production of high brightness γ rays through backscattering of laser photons on high-energy electrons, Physical Review Special Topics-Accelerators and Beams, **6**, 091001 (**2003**).
15. M. Fukuda et al., Polarimetry of Short-Pulse Gamma Rays Produced through Inverse Compton Scattering of Circularly Polarized Laser Beams, Physical Review Letters **Vol.91**, No.16, 164801-1(**2003**).
16. T. Omori et al., Efficient Propagation of Polarization from Laser Photons to Positrons through Compton Scattering and Electron-Positron Pair Creation, Phys. Rev. Lett. **96**, 114801(**2006**).
17. E. Bulyak et al., Beam dynamics in Compton ring Gamma sources Phys. Rev. ST Accel. Beams9, 094001(**2006**).
18. F. Zomer et al., Polarization induced instabilities in external four-mirror Fabry-Perot cavities. Appl. Opt., 48:6651–6661, 2009.

19. Y. Honda et al., Stabilization of a non-planar optical cavity using its polarization property. Optics Communications, 282(15):3108 – 3112, 2009.
20. Sakaue K. et al., Observation of pulsed X-ray trains produced by laser-electron Compton scatterings. Rev. Sci. Instrum., 80:123304–123310, 2009.
21. ThomX collaboration. ThomX CDR. IN2P3, IN2P3-00448278, 2010.
22. A. Variola et al., Luminosity optimization schemes in Compton experiments based on Fabry-Perot optical resonators. Phys. Rev. ST AB, 14:031001, 2011.
23. S. Miyoshi et al., Photon generation by laser-Compton scattering at the KEK-ATF. Nucl. Instrum. Meth., A623:576–578, 2010.
24. T. Akagi et al., in preparation.
25. P. Bambade et al., Present status and first results of the final focus beam line at the KEK accelerator test facility. Phys. Rev. ST Accel. Beams, 13(4):042801, Apr 2010.
26. J. Bonis et al., Non-planar four-mirror optical cavity for high intensity gamma ray flux production by pulsed laser beam Compton scattering off GeV-electrons, arXiv:1111.5833v2 [physics.acc-ph] 22 Dec 2011.
27. T. Akagi et al., Production of gamma rays by pulsed laser beam Compton scattering off GeV-electrons using a non-planar four-mirror optical cavity, arXiv:1111.5834v2 [physics.acc-ph] 4 Jan 2012.

COMPTON RING WITH LASER RADIATIVE COOLING

E. BULYAK*

NSC KIPT,
Kharkov, 61108, Ukraine
* E-mail: bulyak@kipt.kharkov.ua

J. URAKAWA

KEK,
Tsukuba, Japan
E-mail: junji.urakawa@kek.jp

F. ZIMMERMANN

CERN, Geneva, Switzerland
E-mail: Frank.Zimmermann@cern.ch

Proposed is an enhancement of laser radiative cooling by utilizing laser pulses of small spatial and temporal dimensions, which interact only with a fraction of an electron bunch circulating in a storage ring. The dynamics of such electron bunch when laser photons scatter off the electrons at a collision point placed in a section with nonzero dispersion is studied. In this case of 'asymmetric cooling', the stationary energy spread is much smaller than under conditions of regular scattering where the laser spot size is larger than the electron beam; and the synchrotron oscillations are damped faster. Results of extensive simulations are presented for the performance optimization of Compton gamma-ray sources and damping rings.

Keywords: Radiative cooling, Compton sources

1. Introduction

Storage rings as sources of Compton radiation have many advantages such as high average current of relativistic electrons, low energy losses since the electrons can be circulating during long time, etc.

At the same time the Compton storage rings suffer from large spread of electrons' energy which limits their stability and require special measures and thus expenses, to keep the high-spread bunches circulating stable.

In addition, large spread at nonzero crossing angle (non head-on) decreases yield of gammas: the main figure of merit due to enlarging the bunch length. Also it require a high rf-voltage to provide sufficient energy acceptance of the ring.

The report describes a novel method of decreasing the energy spread and increasing the decrement of the radiative damping. This work is in continuation of our works published in.[1-3]

2. Asymmetric 'Fast' Cooling

Cooling of initially hot bunches is a thermodynamical process of decreasing entropy. (A fraction of entropy is outgoing with the radiation.) For the storage rings this process requires 'three many':

- many betatron oscillations (at collision point);
- many synchrotron oscillations;
- many photons scattered by each electron in average.

In both a theoretical and a digital simulating models for asymmetric cooling, the collision point (cp) is located in a dispersive region of the ring orbit. The vertical cross-section of collision considered is presented in Fig.1. Crossing of the electron bunches and laser pulses is in the horizontal plane perpendicular to the picture plane. In the theoretical model the laser field overlaps electrons with positive vertical coordinate at cp, $z \geq 0$. In the simulation, the laser field is resembled by an array of the individual laser pulses confined in the optical resonators (see Fig.1).

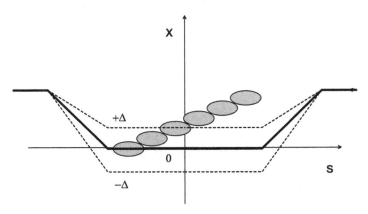

Fig. 1. Trajectories at CP with positive $(+\Delta)$, negative $(-\Delta)$, and zero deviation of the electron energy.

Evolution of the (squared) energy spread, S, and the emittance in direction of the dispersion (vertical in our simulation models) reduced to set of difference equations (1), where time τ is counted in the inverse frequency of scattering events – or number of average scattering.

With β_{cp} being the beta–function magnitude at the collision point (cp), $\varepsilon = \epsilon/\beta_{cp}$ the normalized emittance, $S = \langle p^2 \rangle$ rms spread [$p = (\gamma - \gamma_s)/\gamma_s$, γ_s Lorentz factor of the synchronous electron], $g = D/\beta_{cp}$ the normalized dispersion at cp, changes per the average interaction are:

$$\frac{\Delta\varepsilon}{\Delta\tau} = -\frac{b}{2}\varepsilon + bg\sqrt{2\varepsilon}\,F_s\,(G) + \frac{3b^2}{80\gamma^2}\left(1 + \frac{14}{3}g^2\gamma^2\right)\;; \qquad (1)$$

$$\frac{\Delta S}{\Delta\tau} = -bS - b\sqrt{2S}\,F_c\,(G) + \frac{7b^2}{40}\,.$$

where $b \approx 4\gamma_s\gamma_{las}$ is the maximal recoil undergone by the electron scattered off the laser photon (γ_{las} equivalent Lorentz factor of the laser photons)); $G \equiv g\sqrt{S/\varepsilon}$. Functions $F_s\,(G)$ and $F_c\,(G)$ are averages

$$F_s(G) = \frac{1}{4\pi^2}\iint_0^{2\pi} \mathrm{d}\psi\mathrm{d}\theta\,\sin\theta\,\mathrm{H}(\sin\theta + G\cos\psi)\;;$$

$$F_c(G) = \frac{1}{4\pi^2}\iint_0^{2\pi} \mathrm{d}\psi\mathrm{d}\theta\,\cos\psi\,\mathrm{H}(\sin\theta + G\cos\psi)\,.$$

with $\mathrm{H}(x)$ being the Heaviside step function.

Examples of numerical integration the set (1) are presented in Fig.2.

2.1. *Analysis*

As it follows from the equations (1), the asymmetric laser field induces additional nonlinear damping in synchrotron oscillations and simultaneous excitation in the betatron oscillations (vice versa for negative dispersion). The process is controlled by ratio of the dispersion magnitude at cp to the beta–function magnitude at this point of the orbit.

Steady–state values of the spread and the emittance are as follow (cf.[4] for dispersionless cp):

- Spread $\langle p^2 \rangle$
    - No dispersion: $\langle p^2 \rangle_* = 7\gamma_s\gamma_{las}/10$
    - Positive dispersion: $\langle p^2 \rangle < \langle p^2 \rangle_*$
- Emittance ε
    - No dispersion: $\varepsilon_* \approx 3\gamma_{las}/10\gamma_s$
    - Positive dispersion: $\varepsilon > \varepsilon_*$, exponential growth.

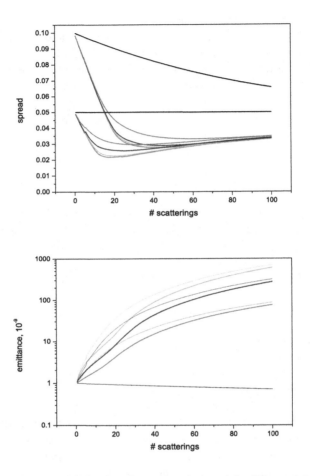

Fig. 2. Kinetics of spread (top) and emittance (bottom) for different initial spreads and different dispersion values. Black curves represent dispersionless case, $g = 0$; pink, blue, green and red ones represent $g = (1, 2, 3, 6) \times 10^{-3}$, resp. $E_e = 0.8\,\text{GeV}$, $E_{\text{las}} = 1.16\,\text{eV}$. Thin curves in the emittance evolution represent the larger initial spread.

As it can be seen from a concrete case in Fig.2, the steady-state 'asymmetric' spread is significantly smaller than the regular one. (The emittance is larger but it can be controlled by proper choice of the beta function and the dispersion.)

Regarding temporal evolution of the spread and emittance, which obeys the classical exponential rule in the no-dispersion cp case:

- Damping of the transversal emittance (initial ε_0, steady ε_*)

$$\varepsilon(\tau) - \varepsilon_* = (\varepsilon_0 - \varepsilon_*)\, e^{-b\tau/2}$$

- Damping of the squared spread – two times faster:

$$\langle p^2 \rangle - \langle p^2 \rangle_* = (\langle p^2 \rangle_0 - \langle p^2 \rangle_*)\, e^{-b\tau}$$

It takes much less time for cooling the initial spread in the asymmetric case than in the symmetric one, also this damping process is not exponential.

3. Employment of Asymmetric Cooling in the Sources of Polarized Positrons

Asymmetric 'fast' cooling would enhance performance of the polarized positron sources.

3.1. Sources of polarized gammas based on Compton storage rings

Asymmetric cooling enables to reduce steady–state energy spread and therefore enhance the yield of gammas due to smaller length of the electron bunches. Also stable circulation of electrons may be obtained at smaller rf voltage, see.[1]

3.2. Damping rings

Asymmetric cooling may be used in the damping rings where positrons are cooling down. For example, every bunch of positrons in an ILC damping ring (population 2×10^{10} positrons at $5\,\mathrm{GeV}$) carries about $16\,\mathrm{J}$ of energy. For $5 \times e$–fold decrease of the spread it is necessary to emit out $80\,\mathrm{J}$ of energy and simultaneously restore it from the rf system.

Damping rings with the fast cooling may operate at sufficiently lower energy of positrons, e.g. $0.7\ldots0.8\,\mathrm{GeV}$. Along with shorter cooling time it would result in more than order of magnitude lower consumption of the rf power.

3.3. Challenge of LHeC: tri-ring scheme

Utter challenge for the damping ring arises from LHeC project.[3] In this project it is required to cool down a continuous $6\,\mathrm{mA}$ positron beam (the

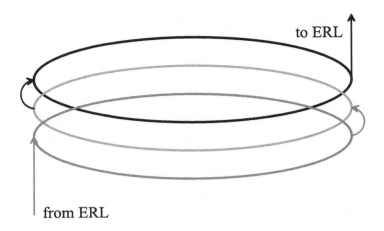

to ERL

from ERL

Fig. 3.   Tri-ring scheme

recycled beam and/or a new beam from a source). Possible solution of this problem is the tri–ring scheme illustrated in Fig. 3.

The operation cycle of the system is as follows:

- The basic cycle lasts N turns.
  - N–turn injection from ERL into the accumulating ring (bottom).
  - N–turn cooling in the cooling ring (middle); fast laser cooling may be employed here.
  - N–turn slow extraction from the extracting ring (top) into the ERL.
- One–turn transfer from the cooling ring into the extracting ring.
- One–turn transfer from the accumulating ring into the cooling ring.

The average current in the cooling ring is $N \times$ average ERL current. The number of turns of the main cycle is limited by the efficiency of multiturn injection and the maximum current which can be stored (and cooled) in the cooling ring.

Laser cooling may generate a new low-emittance positron beams to compensate for losses and emittance growth of the recycled beam.

Reusing and/or cooling of positrons relaxes the requirements for all types of positron source discussed in the following. The cooling period is limited by the maximal stored current in the ring and by the multiturn injection. Fast laser cooling may be employed for compensating positron emittance growth when reusing positrons or to compensate losses (without a dedicated high-current positron source). The slow extraction process is also able to further reduce the energy spread (chromatic extraction) or, alternatively, the transverse emittance (resonant extraction).

4. Summary and Outlook

The asymmetric laser radiative cooling allows reducing one of the most performance-limiting parameters of the Compton-ring gamma sources, namely the energy spread of the circulating bunches, as well as a significant decrease in the cooling period for laser-based damping rings.

The construction of a low-energy damping ring with asymmetric laser cooling appears feasible.

The fast cooling through a localized laser field relaxes the requirements on other parameters for ring-based sources of gamma-ray beams, such as energy acceptance and required RF voltage.[5] It also eases the stacking in a positron accumulator ring for a Compton positron source.

The increase of the transverse emittance due to asymmetric cooling may be tolerable since in laser-dominated rings the 'symmetric' steady-state emittance is very small.

References

1. E. Bulyak, J. Urakawa and F. Zimmermann, Asymmetric laser radiant cooling in storage rings (report mop064), in *PAC 2011, New York, U.S.A., March 28 – April 1, 2011 http://www.bnl.gov/pac11/*, 2011.
2. E. Bulyak, J. Urakawa and F. Zimmermann, Fast cooling of bunches in Compton storage rings, in *IPAC 2011, San Sebastian, Spain, Sept, 2011 http://www.jacow.org/*, rep. *WEPZ011*, 2011.
3. E. Bulyak, Tri–ring scheme, in *Proc. Brainstorming meeting for a LHeC positron source, CERN (to be published)*, 2011.
4. E. Bulyak, P. Gladkikh, L. Rinolfi, T. Omori and J. Urakawa, Beam dynamics in Compton rings with laser cooling (report tupd093), in *IPAC 2010, Kyoto, Japan, 23-28 May 2010*, 2010.
5. E. Bulyak, P. Gladkikh, M. Kuriki, T. Omori, J. Urakawa and A. Variola, Compton gamma-sources: Beam dynamics and performance, in *PosiPol07, Orsay, France*, 2007.

BEPCII POSITRON SOURCE

PEI GUOXI, SUN YAOLIN, LIU JINTONG, CHI YUNLONG, LIU YUNCHENG,
LIU NIANZONG

Institute of High Energy Physics, IHEP, Beijing, P. O. Box 918, 100049, China

BEPCII- an upgrade project of the Beijing Electron Positron Collider (BEPC) is a factory type of e^+e^- collider. The fundamental requirements for its injector linac are the beam energy of 1.89GeV for on-energy injection and a 40mA positron beam current at the linac end with a low beam emittance of 1.6μm and a low energy spread of ±0.5% so as to guarantee a higher injection rate (≥50mA/min) to the storage ring. Since the positron flux is proportional to the primary electron beam power on the target, we will increase the electron gun current from 4A to 10A by using a new electron gun system and increase the primary electron energy from 120MeV to 240MeV. The positron source itself is an extremely important system for producing more positrons, including a positron converter target chamber, a 12kA flux modulator, the 7-m focusing module with DC power supplies and the support. The new positron production linac from the electron gun to the positron source has been installed into the tunnel. In what follows, we will emphasize the positron source design, manufacture and tests.

1. Introduction

BEPCII is the second phased construction of BEPC, working in the Tau-Charm energy region (2-5 GeV). The design peak luminosity of $1\times10^{33}\,cm^{-2}s^{-1}$ at 1.89GeV is two orders of magnitude higher than the present BEPC. To keep a higher average luminosity, on-energy injection with a high injection rate of > 50 mA/min for e^+ beam is necessary, which is ten times the present value. The measures[1] we're going to take include: 1) increase the repetition rate from 12.5Hz to 50Hz; 2) increase the bombarding electron energy from 140MeV to 240MeV; 3) increase bombarding electron current from 2.5A to 6A and 4) manufacture a new positron source to increase the positron yield at the linac end from 1.4 to 2.7% $e^+/(e^-\times GeV)$. Table 1 shows the main parameters of the BEPCII positron source.

Table 1 The main parameters of the BEPCII e+ source

Positron Energy	1.89	GeV
Positron current	37	mA
Energy spread	$\leq\pm0.5$	%
Beam emittance	1.6	μm
Positron source System Target (W)	8	mm
e⁻ energy on target	240	MeV
e⁻ current on target	6	A
Peak field of FLUX	4.5	T
DC field of solenoid	0.5	T
e⁺ yield at solenoid end	0.043	e⁺/e⁻·GeV
e⁺ yield at the linac end	0.026	e⁺/e⁻·GeV

2. Positron converter assembly

The BEPCII positron source[2] is a conventional source. Electrons are accelerated to 240MeV in the linac, and focused to a 3-mm to 5-mm-diameter spot on a tungsten target. The target itself is a 10-mm diameter, 8-mm thick disk. The disk is copper plated, cast in sterling silver with its cooling tubes, and post-machined to size. Simulation shows the mechanical and thermal tensile strengths are not a problem for about 500W electron beam power. The target arm assembly is suspended and supported in the vacuum chamber by the target housing mounted outside the chamber. A bellows is used to feed the target into the vacuum chamber and provide a flexible pivot. The actuation system is an eccentric axletree. With a stepping motor, it can easily move the target in and out of the beam line.

The pair-produced positrons out of target have divergent angles, a broad energy spectrum, and a much larger emittance than that of the initial electron beam; thus a matching device is first needed to transfer an "erect ellipse" transverse emittance of the positron beam to a "horizontal ellipse" acceptance of the downstream system. We use SLAC type matching device, a flux concentrator[3] (FLUX). It is a 12-turn, 10cm long copper coil with a cylindrical outside radius of 53mm and a conical inside radius growing from 3.5mm to 26mm. The 0.2mm gaps between the individual windings are manufactured by electric discharge machining out of one copper block. Excitation current and water-cooling is provided by a hollow rectangular copper conductor brazed to the outside of the coil (also 12 turns). Figure 1 shows the picture of the flux concentrator and its magnetic field profiles. The dotted line is the OPERA simulated result and the solid line is the measured result [4], both are well matched.

A high DC magnetic field of tapered field solenoid (TFS) is required at the target downstream face to capture the low energy positrons. POISSON computer code is used to optimize the flux return iron geometry and coil placement around the target chamber. At a maximum current of 750A, the TFS generates a 12kG DC field at the target exit face. A pair of steering coils is considered to compensate the stray field. The vacuum chamber is contained within the TFS. It is a custom design to maintain structural rigidity, provide the maximum pumping speed to the target and flux concentrator area, accommodate the pole tip section of the TFS flux return iron, mate up to the existing RF structure section, and allow the target and flux concentrator to be removed independently. The downstream vacuum flange of the chamber is quick-disconnect flange. A custom aluminium seal between the chamber flange and the mating flange of the accelerator structure downstream is used. A single convolution bellows between the chamber and the flange provides enough compliance to take up any slight angular misalignments of the two mating flanges while still being rigid enough to support the flange without vertical restraint. Figure 2 is the picture of the BEPCII positron converter assembly.

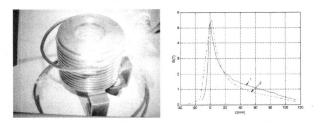

Figure 1   Picture of the e^+ FLUX and its field profiles

Figure 2   Picture of the e^+ converter assembly

3. Positron flux modulator

The flux modulator for the BEPCII positron source provides 12kA in a 5μs sinusoidal half wave current at 50pps to produce an adiabatic magnetic field profile with the peak of 4.5T at the flux entrance face. The modulator uses two CX1536 thyratrons in a switching network, and provides reliable operation with acceptable thyratron lifetime.

There are two such basic modulators, one is the SLAC flux modulator[5] and the other one is the PLS design for the injection kicker[6]. Their basic circuits are the same, an RLC resonant circuit. The SLAC design uses 8 forward CX1622 glass thyratrons to generate pulse current and uses 2 inverse CX1622s and water cooled resistor to minimize inverse voltage and uses resonant charging DC power supply. The PLS design only uses 2 CX1536 thyratrons to generate pulse current and uses RC auxiliary circuit and energy dump circuit to minimize inverse voltage. The BEPCII flux modulator uses PLS design because it is much easier to synchronize trigger time. Also, CCDS charging power supply is more compact and more efficient than the resonant charging DC power supply.

A schematic circuit diagram of the BEPCII flux concentrator modulator is shown in Figure 3. The charge assembly of the modulator mainly consist of a CCDS power supply, a charging resistor, the charging diodes, and main capacitors. The discharge assembly is two thyratrons, inverse energy dump assemblies, and transient suppression assemblies. The reduction of system inverse voltage is a design issue to make a reliable operation and meet the pulse requirements. The peak inverse voltage, including spike must not exceed 10kV for the first 25 µs after the anode pulse, because this can cause reverse conduction in the thyratron and therefore damage the tube. The other issue is the inductance reduction, because the flux concentrator modulator design requires a high current pulse into an inductive load. This is accomplished by building wide current paths while minimizing the current loops. The modulator has 15m long multi cables, and they are connected in parallel at the load. Creating 20 parallel paths reduces the system inductance, thus increasing the system reliability.

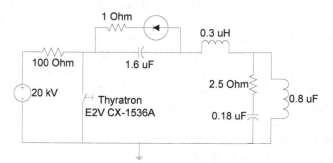

Figure 3 Circuit diagram of the BEPCII flux concentrator modulator

The new modulator was installed in the klystron gallery in the summer shutdown of 2003, and first operated for BEPC matching solenoid at 6kA, 12.5pps. Since the new positron source installed in the tunnel at the October 2004, we've tested the modulator with the flux concentrator at 50pps, 12kA. Figure 4 is the pulse current of flux concentrator, the pulse bottom width is about 8µs, and because of the mismatch between the impedance of the load and the pulse cables, reflections occur up and down the cables adding an oscillation on top of the load current. The saturable inductors in series with the thyratron can help minimize these oscillations by initially slowing the rate of rise of voltage across the load.

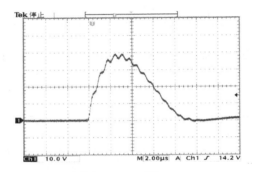

Figure 4 Pulse current of flux concentrator

4. Focusing module and support

The positrons out of target have larger divergent angles. And an intensive longitudinal magnetic field of 0.5T which provides a maximum transverse acceptance of 0.31π(MeV/c)-cm, is required to confine the positrons during capturing and accelerating. POISSON has been run to optimize the coil arrangements and the flux return iron geometry. At a DC driving current of about 400A, the seven 1-meter long focusing modules can provide 0.5T magnetic field with a drop of less than 5% in the gap between two modules. Between each module and RF structure, there is a pair of steering coils to guide the positron beam. The positrons out of the focusing modules are about 100MeV, which makes good match into the downstream quadrupole focusing system.

The hollow square conductor is used to wrap the pancakes with insulating tape, and then the coils are encased and epoxyed. The electrical connection is in series, and the cooling circuits connected in parallel. The epoxyed pancakes are assembled and aligned in the case which is fabricated from low carbon steel.

The pancake's second epoxy is conducted to form the focusing module. Since the focusing modules are heavy, each of about 2 tons, the support structures[7] are newly designed to replace the existing girders as shown in Figure 5. The focusing modules are fastened to moveable platform supported on the rails oriented parallel to beam axis. A little lateral motion is possible for alignment. In order to install the modules on the RF structures, the output coupler is specially designed with a short flange and the cooling tube joints are moved to the input coupler end.

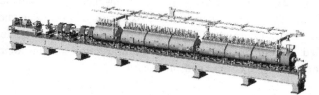

Figure 5a   The positron source 3D model

Figure 5b   Picture of the positron source

5. Interlocks

The whole positron source is comprised of several interlocks, many of which are attributed to the flux modulator and DC power supplies. Besides the personnel safety interlocks, there are also interlocks for equipment protection, including a water flow interlock for the water cooled resistors and power supplies, an over-current interlock that monitors the input current of the power supply, a temperature sensor for the water flow, a vacuum interlock for the power supply, and a crowbar interlock to protect the modulator from excessive charging current and charging voltage.

6. Conclusion

The new BEPCII positron source has been designed and fabricated since 2002, and the whole system installed into the tunnel1 in five months from May 1st of 2004. Each system has been tested with satisfactory results. Now the BEPC linac is running for synchrotron radiation use up to the end of June in 2005. From then on, the positron linac commissioning has been conducted, and the first positron beam of higher than 40 mA was obtained at the linac end on March 19, 2005. A positron current of higher than 60mA at the positron solenoid end was obtained. It confirms that the design, construction and installation of this new positron source are correct and successful.

References

1. G.X. Pei, et al., Design Report of the BEPCII Injector Linac, IHEP-BEPCII-SB-03-02, November 2003.
2. W.P. Gou, G.X. Pei, The Physical Design of the BEPCII Positron Source, High Energy Phys. and Nucl. Phys., 2002, 26(3): 279--285.
3. A.V. Kulikov, et al., SLC Positron Source Pulsed Flux Concentrator, Proceedings of 1991 IEEE Particle Accelerator Conference, May 6-9, 1991, San Francisco, P. 2005, SLAC-PUB-5473 (1991).
4. J.T. Liu, et al., Development of the BEPCII Positron Source Flux Concentrator, to be published in High Energy Phys. and Nucl. Phys.
5. J. de Lamare, et al., SLC Positron Source Flux Concentrator Modulator, Proceedings of 1991 IEEE Particle Accelerator Conference, May 6-9, 1991, San Francisco, P. 3138, SLAC-PUB-5472 (1991).
6. S.H. Nam, et al., Injection kicker modulator in 2Gev pohang light source, Proceedings of 1998 Twenty third international power modulator symposium.
7. Y.L. Sun, BEPCII Positron Source Mechanical Design Considerations, inner report.

RESEARCH OF COMPACT X-RAY SOURCE BASED ON THOMSON SCATTERING AT TSINGHUA UNIVERSITY

YINGCHAO DU(杜应超), LIXIN YAN(颜立新), JIANFEI HUA(华剑飞), QIANG DU (杜强), HOUJUN QIAN(钱厚俊), CHEN LI(李晨), HAISHENG XU(许海生), WENHUI HUANG(黄文会), HUAIBI CHEN(陈怀璧), CHUANXIANG TANG(唐传祥)

Accelerator laboratory, department of engineering physics, Tsinghua University, Beijing, 100084, China

Key laboratory of Particle and Radiation Imaging (Tsinghua University), Ministry of Education, Beijing, 100084, China

Key laboratory of High Energy Radiation Imaging Fundamental Science for National Defense, Beijing, 100084, China

Recently, good X-ray pulses have been generated successfully via Thomson scattering between the 47MeV electron and 800nm TW laser at Tsinghua Thomson scattering X-ray source. In this paper, we report the preliminarily results of the experiments.

1. Introduction

Thomson scattering X-ray source is an excellent compact X-ray source which can provide ultra-short, high peak brightness, monochromatic, tunable polarized hard X-ray pulse with table size accelerator and TW laser system. It has wildly application potential in various scientific research fields as well as in medical, technology and industrial fields [1]. Many institutes [2-8], including accelerator laboratory in Tsinghua University [9-13], have studied and developed such X-ray source in the last decade.

The scheme of the Tsinghua Thomson scattering X-ray source (TTX) is shown in Figure 1. It consists of a femtosecond 20TW Ti:Sapphire laser system and a 50MeV linac with photocathode RF gun. The laser system generates both the 266nm UV pulse for photocathode and the 800nm IR pulse for scattering interaction. The two pulses are derived from one 79.3MHz Ti:Sapphire oscillator in order to reduce the time jitter between the electron beam and the IR pulse. The linac system consists of a BNL/KEK/SHI type 1.6 cell S-band photocathode RF gun, a 3m S-band SLAC type traveling wave accelerating section, a four dipole magnet compressor to compress the electron bunch to below 1ps, and two X-band harmonic structures, enabling flexible manipulation of phase space of electron pulses, generates 40~50MeV ultra-short high brightness electron pulse for scattering interaction. The laser system is synchronized with the RF system through a timing circuit, with a timing jitter no greater than 0.5ps. The electron and IR laser pulses collide in the interaction chamber with geometries ranging from 90 deg to 180 deg, and generate 20~50keV X-ray pulse with fluxes from 10^6 to 10^8 photons/pulse and pulse durations from 200fs to 1ps. This X-ray source will be severed as the tools for X-ray phase contrast imaging, hard x-ray polarimeter calibration, ultra-fast pump-probe experiments, and so on.

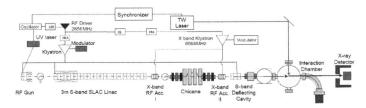

Figure 1. Scheme of TTX

Figure 2. The accelerator of TTX

Figure 3. Laser system of TTX

Now we have finished the installation of the systems expect the X-band RF system and the magnetic bunch compressor, the linac and the laser system are shown in figures 2 and 3.

2. Description of the experiment

The head on colliding experiment has been demonstrated successfully at TTX in 2011. The layout of the interaction chamber region is shown in figure 4. The electron beam is focused by a set of quadrupole magnets with a magnetic field gradient of up to 12T/m. The IR laser is focused by a 5 inch focal length, 90degree off-axis parabolic mirror and head on scattered by the electron beam, and then guided to the laser dump by another parabolic mirror placed at upstream of the beam line. There is a 4mm diameter hole on the center of the parabolic mirrors to enable electron beam and generated X-ray photons to pass through. The mirrors in the chamber are remote controlled by the step motor which allows for control of the transverse alignment of the laser the interaction point. A 45degree bend dipole magnet is placed after the interaction chamber to bend the electron beam and separate the electron beam from the scattered X-ray, which propagate in the same direction as the electrons. A 100um thick Ti window is installed at the beam line exit as the X-ray window to separate the beam line vacuum from the atmosphere.

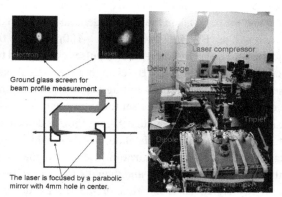

Figure 4. Layout of the interaction region of TTX

There are two types of X-ray detectors for generated X-ray diagnosis in the experiment. One is the removable circular micro-channel plate detector (MCP) which is placed after the dipole to measure the X-ray flux, including the bremsstrahlung background and scattered X-ray. The MCP also is served as the monitor for the preliminarily longitudinal alignment between the electron and laser. Another is the profile monitor which includes a CsI crystal, an imaging CCD camera. We use this detector to measure the X-ray transverse distribution.

The transverse and longitudinal alignments between the electron and TW laser beam are the key points to obtain good X-ray signal during the experiment. The transverse alignment of the two focal spots is performed with the aid of a removable thin ground glass at the center of the interaction chamber. This ground glass is mounted on a vacuum feed through, with its faces oriented vertically normal to the beam line and the horizontally at 45degree to the beam line. When the electron strike the glass, the Cherenkov light is produced and can be imaged into the CCD camera. The laser beam diffuse reflects from the depolished surface, and can also be imaged by the CCD camera. The typical beam profile of the electron and laser are shown in figure 4. Because it is more critical that the electron beam pass through the small hole in the mirrors without any beam loss which causes the heavy bremsstrahlung X-ray background, generally the electron beam is focused and fine tuned with the focal magnets and steering magnets on the upstream beam line, and then the laser is aligned to the electron focus position. After the spatial alignment, the temporal synchronization is preformed then. There are three steps to the initial synchronization. First, the focused laser and the electron beam strike the edge of the glass at the same time, the MCP detector is used to monitor the harmonic photos from the IR and the bremsstrahlung X-rays from the electron beam. The charge and IR energy is reduced to avoid the saturation of MCP and the damage on the screen. Generally, we can detect the two similar signals as ~400ps FWHM pulses with ~100ps rising timing by an oscilloscope, and the time delay between the two pulses correspond to the arriving timing difference of electron and laser. By changing the optical path of the UV laser and IR laser, the electron and laser timing are brought to within about 100 ps. Second, the charge and IR energy are set to the normal value, then the delay line of the IR laser is scanned and the scatted X-ray signal is detected by the MCP. The adjustment range of the delay line is about +/- 3cm, which is large enough after the first stage of temporal synchronization. Generally, the X-ray signal can be detected by the MCP during the scan, and then the timing can be optimized by maximizing the X-ray signal as a function of the timing delay between the electron and IR laser.

3. Experimental results

Typical background and scattered X-ray signal detected by the MCP from the oscilloscope is shown in figure 5. The red line is the background signal with the electron beam on and IR laser off, there are two small peaks on the background signal. The first one comes from the beam loss at the upstream beam line, especially from the two parabolic mirrors. The second peak which is later about 3ns than the first comes from the beam dump following the dipole magnet. The blue line is scattered X-ray signal while the electron and IR laser are on at the same time. The scattered X-ray signal and background signal from the beam loss at the upstream beam line are synchronized. This is why we can use the background signal for longitudinal synchronization.

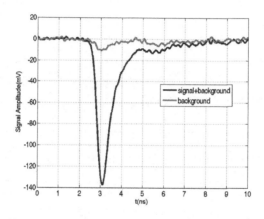

Figure 5. Typical X-ray signal from MCP. Red line is the background signal with electron on only. The blue line is scattered X-ray signal with electron and IR laser on.

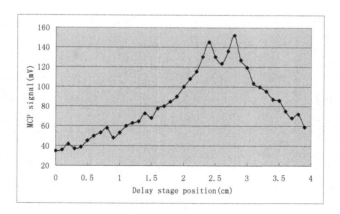

Figure 6. X-ray signal amplitude as a function of the delay line position.

The signal amplitudes as a function of the IR laser timing delay is shown in figure 6. During the delay line scan, it seems that there are two positions to generate maximum X-ray flux. The reason is not clear yet. The initial suspicion is the IR laser and electron are focused in the different longitudinal position, and it will be verified in the next experiment. The signal as a function of the IR laser energy is shown in figure 7. Linearity is good between the IR laser power and intensity of the X-ray. The maximum number of photons is estimated with the signal to be 1.2x10^5/pulse with 300mJ laser power.

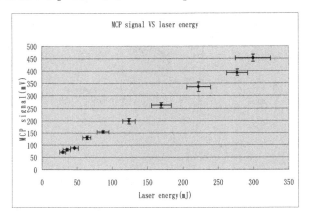

Figure 7. X-ray signal amplitude as a function of the IR laser power

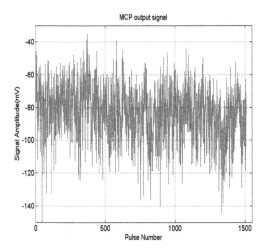

Figure 8. long time X-ray intensity fluctuation.

Figure 8 gives the intensity fluctuation with 1600 pulses. The rms jitter is about 18%. The contribution of the beam charge and laser power fluctuation is about 6% and 7%, respectively. The residual fluctuation may come from the synchronization jitter and transverse position jitter. Detail investigation with simulation and experiment is ongoing now.

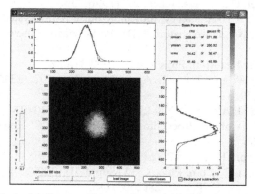

Figure 9. Single shot X-ray profile measurement with CsI screen and EMCCD.

The profile of the X-ray is measured with a CsI screen and EMCCD camera. The single shot image is shown in figure 9. The X-ray imaging study will be start soon.

4. Outlook

We have generated high signal-to-noise X-ray pulse at Tsinghua Thomson scattering X-ray source successfully. The parameters of the X-ray is preliminarily measured and optimized. In the following experiments, we will increase the stability of the machines to get more stable X-ray. Detail X-ray parameters including flux, the spectrum, polarization, transverse distribution, and so on, will also be studied. And the preliminary applications will be start soon.

Acknowledgement

This work is supported by the National Natural Science Foundation of China (NSFC) under Grant Nos. 10735050, 10805031, 10975088 and 10875070 and by the National Basic Research Program of China (973 Program) under Grant No. 2007CB815102.

References

1. Scientific Needs for Future X-ray Source in the US – A white paper, SLAC-R-910, 2008.
2. K.J.Kim, et al. Nucl. Instr. and Meth. A 341(1994)351.
3. A.Ting, et al. Nucl. Instr. and Meth. A375(1996)ABS68.
4. R.W.Schoenlein, et al.,Science, 274(1996)236.
5. D.J.Gibson. A High-energy, ultrashort-pulse X-ray system for the dynamic study of heavy dense materials[D]. Ph.D. thesis, University of California Davis, 2004
6. C.Brau. Nuc Inst Meth Phys Res A. 1992,318:38
7. H. Ikeura-Sekiguchi, et al. Appl. Phys. Lett. 92, 131107 (2008); doi: 10.1063/1.2903148
8. P. Oliva, et al. APPL. PHYS. LETT. 97, 134104(2010)
9. C.X. Tang et al. in Proceedings of LINAC 2006. Knoxville, Tennessee USA, p. 256
10. Y.C. Du, et al., HEP & NP 32 (2008) 75.
11. C.X. Tang, et al., Chin. Phys. C 33 (Suppl. II) (2009) 146.
12. X.Zh. He, et al., HEP& NP 28(2004)446.
13. Y.Ch. Du, et al., Nucl. Instr. and Meth. A637(2011)S168-S171.